Analyses for Hormonal Substances in Food-producing Animals

RSC Food Analysis Monographs

Series Editor:
P.S. Belton, *School of Chemical Sciences, University of East Anglia, Norwich, UK*

Titles in the Series:
1: Dietary Fibre Analysis
2: Chromatography and Capillary Electrophoresis in Food Analysis
3: Quality in the Food Analysis Laboratory
4: Mass Spectrometry of Natural Substances in Food
5: The Maillard Reaction
6: Extraction of Organic Analytes from Foods: A Manual of Methods
7: Trace Element Analysis of Food and Diet
8: Analyses for Hormonal Substances in Food-producing Animals

How to obtain future titles on publication:
A standing order plan is available for this series. A standing order will bring delivery of each new volume immediately on publication.

For further information please contact:
Book Sales Department, Royal Society of Chemistry,
Thomas Graham House, Science Park, Milton Road, Cambridge,
CB4 0WF, UK
Telephone: +44 (0)1223 432360, Fax: +44 (0)1223 420247, Email: books@rsc.org
Visit our website at http://www.rsc.org/Shop/Books/

Analyses for Hormonal Substances in Food-producing Animals

Edited by

Jack F. Kay
Department of Statistics and Modelling Science, University of Strathclyde, Glasgow, UK

RSC Publishing

RSC Food Analysis Monographs No. 8

ISBN: 978-0-85404-198-5
ISSN: 1757-7098

A catalogue record for this book is available from the British Library

Published by The Royal Society of Chemistry,
Thomas Graham House, Science Park, Milton Road,
Cambridge CB4 0WF, UK

Registered Charity Number 207890

For further information see our web site at www.rsc.org

Dedication

To Carol, my love and my anchor, whose selfless love and devotion to her family and friends know no limit.

אשת חיל מי ימצא, ורחק מפנינים מכרה.

בטח בה לב בעלה, ושלל לא יחסר.

גמלתהו טוב ולא רע, כל ימי חייה.

רבות בנות עשו חיל, ואת עלית על־כלנה.

A woman of strength, who can find? Her worth is far beyond pearls.
Her husband's heart trusts in her, and he has no lack of gain.
She brings him good, not harm, all the days of her life.
Many women have excelled, but you surpass them all.
(Proverbs 31: 10–12, 29)

Preface

The genesis of this book can be traced to a meeting held in November 2006 at HFL Limited (now HFL Sport Science), a UK laboratory with significant experience in the detection of veterinary drug residues in biological matrices. The ban on the use of hormones and related compounds for growth promotion in the European Community has been successively increased since the initial ban on stilbenes was introduced in 1981 and now includes β-agonists as well as other substances having a hormonal action such as thyrostats. The original purpose of the meeting at HFL was to review progress on a multinational European Commission supported study to detect the illegal administration of growth-promoting hormones. This has been a significant theme within the United Kingdom government's R&D programme since the early 1990s and is co-ordinated by the Veterinary Medicines Directorate (VMD), an agency of the Department for Environment Food and Rural Affairs (Defra).

The European study built on earlier UK supported studies and there were many delegates present from across the European Union, so this was seen as an ideal opportunity to extend the meeting to review progress in the wider area of hormone and related residue detection in food-producing species. By including this additional discussion, participants were able to gain a wider appreciation of current developments in a number of different countries.

This meeting was a success and was considered by those who spoke to me afterwards as being very useful. With this in mind, I thought that more benefit could be gained from the meeting. When the "hormone ban" first arose in the EU, there were only 15 Member States. There are now 27 Member States, many of which do not share the historical background to the issue enjoyed by those who attended our meeting.

This volume attempts to capture the key issues discussed in the above meeting and extends the coverage to ensure that the most recent developments

RSC Food Analysis Monographs No.8
Analyses for Hormonal Substances in Food-producing Animals
Edited by Jack F. Kay

Published by the Royal Society of Chemistry, www.rsc.org

in the area are addressed in a single reference work. However, the subject area discussed has been limited to growth promoting hormones, although for completeness there will be some reference to the wider range of growth promoting substances which have been used around the world either legally or otherwise.

This book is intended as a purely factual account of the background to what has become a significant international issue for both consumer protection and trade. It deals with how science has developed to answer the questions raised. It does not set out to argue the merits of the cases for and against the use of growth promoting hormones in food animal production. The authors of the individual chapters are internationally acknowledged experts in their field and the purpose of this book is to provide a single and definitive source of information setting out the historical position of the trade issue which brought this dispute to public attention, together with an overview of the high quality science generated in response to this to protect consumers. It is my hope that this will be of benefit to scientists working in this area, together with regulators and consumers.

I am particularly indebted to the publishers in extending the deadline for the text of this book, as this allowed the inclusion of relevant discussions at the 18th session of the Codex Committee on Residues of Veterinary Drugs in Food (CCRVDF) in May 2009, which was held in Natal, Brazil. The subject of growth promoting hormones was once again on the agenda for this CCRVDF meeting and it was fitting to make this reference work as up to date as possible.

I would like to thank the VMD for allowing me the opportunity and time to accept an honorary position at the University of Strathclyde, under which I have prepared this book. Thanks are also due to Professor George Gettinby for offering me the position and for all his support in preparing this volume.

Jack F. Kay

Contents

RSC Food Analysis Monographs No.8
Analyses for Hormonal Substances in Food-producing Animals
Edited by Jack F. Kay

Published by the Royal Society of Chemistry, www.rsc.org

CHAPTER 1

The Use of Hormomally Active Substances in Veterinary and Zootechnical Uses – The Continuing Scientific and Regulatory Challenges

LEONARD S. LEVY

Institute of Environment and Health, Cranfield University, Cranfield, Bedfordshire, MK43 OAL, U.K.

1.1 Introduction

"Those who cannot learn from history are doomed to repeat it."
(George Santayana, 1863–1952)
"What experience and history teach is this – that people and governments never have learned anything from history, or acted on principles."
(George Wilhelm Hegel, 1770–1831)

The scientific, regulatory and political debate surrounding the use or banning of hormones or hormone-like substances in the production of meat and meat products and for veterinary use is, for me, the quintessential example of the limitations of all three of these facets of risk assessment/management inputs in

RSC Food Analysis Monographs No.8
Analyses for Hormonal Substances in Food-producing Animals
Edited by Jack F. Kay

Published by the Royal Society of Chemistry, www.rsc.org

producing unequivocal answers. It is thus a salutary lesson to scientists, regulators and policy-makers, politicians and risk assessors to understand all strands of this tangled issue. Whether or not it helps us to make better decisions in the future, for this and other continuing debates in risk assessment, will depend on which of the two above apparently contradictory quotations regarding the utility in understanding the past you subscribe to.

1.1.1 Recent Historical Perspective

A range of hormonally active substances, such as diethylstilboestrol (DES), had been used for growth promotion in cattle and sheep since the early 1950s. This latter use is termed zootechnical as opposed to veterinary or therapeutic use. Concerns about a possible risk of cancer from residues of such substances had been expressed in the early 1970s and eventually the European Community (EC) introduced a ban on the use of DES in 1987 and, in addition, banned the use of all hormonally active substances as growth promoters in food-producing animals in 1988. A similar condition was placed on all countries, so-called "Third Countries", wishing to export meat from such animals to the EC.

The United States and Canada objected to this ban to the World Trade Organization (WTO). As a result, in 1997, the WTO Expert Panel found that the ban was not based on science – for example, on a risk assessment or on relevant international standards.[i]

The European Commission appealed against this ruling. In February 1998, the Appellate Body upheld the WTO Expert Panel's view, in that they found that the ban had been imposed *without credible evidence* to indicate that there were health risks posed by eating hormone-treated meat. As a result, the European Commission was given 15 months to remove the ban or produce a risk assessment.

As a response to this ruling, in early 1998, the European Commission (EC) sponsored 17 research studies to help clarify the findings in the Appellate Body report. These covered toxicological and carcinogenicity aspects, residue analysis, potential abuse and control problems and environmental effects of hormone use.

At the end of 1998, the EC's Scientific Committee on Veterinary Measures relating to Public Health (SCVPH) was asked to carry out an assessment of the risk to human health from the use of the six hormonally active substances, particularly from residues from bovine animals where such substances were administered for growth promotion. These substances were: 17β-oestradiol, testosterone, zeranol, progesterone, trenbolone acetate and melengestrol acetate. In April 1999, the SCVPH produced its first Opinion on the subject.[1]

The SCVPH concluded that the risks from hormone-treated meat were "higher than previously thought". Further, it proposed that there was a significant body of scientific evidence suggesting that 17β-oestradiol should be considered a complete carcinogen. It also concluded, with different standards of evidence, that there were risks to consumers from the other five hormones examined.

[i] The WTO Panel report numbers – WT/DS26/R/USA and WT/DS48R/CAN.

Most importantly, the SCVPH concluded that no threshold concentrations could be defined for the hormones – this precluded the setting of Acceptable Daily Intakes (ADIs) or Maximum Residue Limits (MRLs). However, they were unable to estimate the extent of any risk.

In the UK, the then Minister of Agriculture, Fisheries and Food asked the Veterinary Products Committee (VPC) to assess the evidence in the SCVPH Opinion. The VPC is an independent scientific committee that has the remit to give scientific and veterinary advice on veterinary medicines and other products used for animal production and husbandry to the Veterinary Medicines Directorate (VMD). The VMD has statutory duties in relation to veterinary medicines and products in the UK. The VPC set up a Sub-Group to do this, which reported in October 1999. At the same time, the Safety Working Group of the Committee for Veterinary Medicinal Products (CVMP) – the European Commission's own organisation with responsibility for advising on the safety of veterinary medicines – also examined the SCVPH Opinion.

Following detailed deliberations, the VPC Sub-Group was unable to support the SCVPH's conclusion that the risks associated with eating hormone-treated meat "might be higher than previously thought". The Sub-Group also found that it had sufficient concerns about the scientific reasoning in a number of key areas, to throw serious doubt on the conclusions of the SCVPH. However, the Group identified a number of areas where additional expert evidence should be sought to add to the data and help prevent selective scientific conclusions being drawn in the future.

The CVMP also produced a report in 1999 in response to the SCVPH Opinion (EMEA/CVMP/885/99). The CVMP was unconvinced by the SCVPH data and arguments, and concluded that its (the CVMP's) previous recommendations with regard to the ADIs and MRLs of the five hormones examined were still applicable (17β-oestradiol, altrenogest, progesterone, flugesterone acetate and norgestomet). The CVMP also noted that its conclusions were practically the same as the FAO/WHO Joint Expert Committee on Contaminants and Food Additives.[2]

The UK Government accepted the view of the VPC-that they were unable to support the conclusion of the SCVPH of a "higher risk than previously thought" from eating hormone-treated meat. The UK has, however, always fulfilled its obligations to enforce the EU ban and continues to do so.

In May 2000, the SCVPH produced a review[3] of its Opinion after having examined the reports of both the VPC and CVMP. The SCVPH noted that these two independent evaluation reports showed a high degree of consensus on the possible risks. However, it did not seek to answer the questions raised in the reports, but concluded that they did not provide convincing data and arguments that demanded revision of the SCVPH's previous conclusions. The SCVPH review acknowledged that there were obvious gaps in the present understanding on the hormones in relation to animal metabolism and residue deposition but it anticipated that the EC's research programmes the EC had instigated would provide additional data on these topics.

Following the completion of these 17 studies sponsored by the EC, the SCVPH was asked to review their previous Opinions of 1999 and 2000, the data from the 17 studies and other recent scientific literature from any source. In April 2002, the SCVPH released yet another Opinion.[4] In this, it reconfirmed the views in the previous SCVPH Opinion and concluded that no amendments to these were justified.

In September 2003, the European Parliament and Council of Ministers passed Directive 2003/74/EC.[ii] This Directive put further restrictions on the use of veterinary medicinal products containing oestradiol or its ester-like derivatives. Originally, the intention was for all uses of 17β-oestradiol to be banned and restrictions tightened on other hormones; however, the UK and other Member States expressed concerns about the potential loss of a number of valuable veterinary therapeutic products.

This Directive required that oestradiol and its derivatives should not be used for oestrus induction/synchronisation in cattle, horses, sheep or goats after October 2006. These substances were still allowed to be authorised for the treatment of foetus maceration or mummification and the treatment of pyometra in cattle, or oestrus induction in cattle, horses, sheep or goats. However, the Directive required the European Commission to present a report by October 2005 on the possible alternatives to oestradiol for these therapeutic uses. A new Directive 2008/97 now bans the use of oestradiol in cattle, *etc*. The key conclusions of the SCVPH Opinion are reproduced below:

"The review of the 17 studies launched by the European Commission and a recent scientific literature allows the following conclusions:

- Ultra-sensitive methods to detect residues of hormones in animal tissues have become available, but need further validation.
- Studies on the metabolism of 17β-oestradiol in bovine species indicate the formation of lipoidal esters, disposed particularly in body fat. These lipoidal esters show a high oral bioavailability in rodent experiments. Thus, the consequence of their consumption needs to be considered in a risk assessment.
- Experiments with heifers, one of the major target animal groups for the use of hormones, indicated a dose-dependent increase in residue levels of all hormones, particularly at the implantation sites. Misplaced implants and repeated implanting, which seem to occur frequently, represent a considerable risk that highly contaminated meats could enter the food chain.
- There is also a dose-dependent increase in residue levels following the oral administration of melengestrol acetate at doses exceeding approved levels, with a corresponding increased risk that contaminated meats could enter the food chain.
- Convincing data have been published confirming the mutagenic and genotoxic potential of 17β-oestradiol as a consequence of metabolic

[ii] http://eur-lex.europa.eu/LexUriServ/LexUriServ.do?uri=CELEX:32003L0074:EN:HTML, accessed 29 March 2009.

activation to reactive quinones. *In vitro* experiments indicated that oestrogenic compounds might alter the expression of an array of genes. Considering that endogenous oestrogens also exert these effects, the data highlight the diverse biological effects of this class of hormones.

- No new data regarding testosterone and progesterone relevant to bovine meat or meat products are available. However, it should be emphasised that these natural hormones are used, only in combination with 17β-oestradiol or other oestrogenic compounds in commercial preparations.
- Experiments with zeranol and trenbolone suggested a more complex oxidative metabolism than previously assumed. These data need further clarification as they might influence a risk assessment related to tissue residues of these compounds.
- Zeranol and trenbolone have been tested for their mutagenic and genotoxic potential in various systems with different endpoints. Both compounds exhibited only very weak effects.
- Data on the genotoxicity of melengestrol acetate indicate only weak effects. However, pro-apoptotic effects were noted in some cell-based assays, which were attributed to the impurities in commercial formulation. Further experiments should clarify the toxicological significance of these impurities.
- Model experiments with rabbits treated with zeranol, trenbolone or melengestrol acetate, mirroring their use in bovines, were designed to study the consequences of pre- and perinatal exposure to exogenous hormones. All compounds crossed the placental barrier easily and influenced to varying degrees the development of the foetus, at the doses used in the experiments.
- Epidemiological studies with opposite-sexed twins suggest that the exposure of the female co-twin *in utero* to hormones results in an increased birth weight and consequently an increased adult breast cancer risk.
- Several studies were devoted to the potential impact of the extensive use of hormones on the environment. Convincing data were presented indicating the high stability of trenbolone and melengestrol acetate in the environment, whereas preliminary data were provided on the potential detrimental effects of hormonal compounds in surface water. In conclusion, after re-appraisal of the data from the 17 studies and recent scientific literature, the SCVPH confirms the validity of its previous Opinions (in 1999 and 2000) on the Assessment of Potential Risks to Human Health from Hormone Residues in Bovine Meat and Meat Products, and that no amendments to those Opinions are justified."[4]

Due to this new SCVPH Opinion, in 2002 the VPC was again asked to examine the scientific evidence for a ban on the use of hormones in food-producing animals and to advise on whether therapeutic uses posed any risk to consumers. A new VPC Working Group was formed with the following Terms of Reference:

- "to evaluate the latest Opinion of the Scientific Committee on Veterinary measures relating to Public Health (SCVPH) dated April 2002 and advise on its conclusions and;

- to advise on whether the latest Opinion of the SCVPH, and the research studies on which it is based, addresses the conclusions reached in the report by the VPC Sub-Group published in October 1999."

1.1.2 The Role and Remit of the VPC 2002 Working Group

The Working Group (WG) was set up in November 2002 and consisted of members of the VPC plus a number of additional invited experts required covering the specialisation required to address all the relevant scientific and veterinary areas. It was agreed by the WG that, in addition to a critical evaluation of the SCVPH Opinion and the 17 EC-funded studies, other recent relevant scientific publications would be sought, evaluated and included, based on the knowledge and expertise of WG members. In some cases, only the study reports of some of the 17 studies submitted to the EC were available. In other cases, peer-reviewed publications were available, based on results within some of the study reports.

The WG also agreed that, apart from a scientific evaluation of existing data, it was important to highlight significant gaps in knowledge and areas where uncertainty existed. This was felt to be particularly relevant as the WG was aware that it had no mandate to make, or even propose, on policy in this area; thus, it was crucial for the report to express uncertainties where they existed in the science base which would inform policy makers and other stakeholders.

Another issue of concern within the WG was to ensure that the reader understood the various uses to which hormonal substances had been, or could be, used in bovine meat production and ensuing meat products. Their use as growth promoters may be regarded as zootechnical. This was the original use for which the EU ban was intended. Oestrus induction in cattle and horses is also a zootechnical use whereas other veterinary uses such as the treatment of pyometra are regarded as therapeutic.

It was noted by the WG that, although the toxicological evidence of the substances under discussion would be based on studies on specific laboratory animals, the actual risk assessment to humans would be dependent on the dose used, the time of administration relative to slaughter, the pharmacological formulation and route of application to these meat-producing animals, and information on human consumption; questions of whether risks, however minimal, should be permitted – depending on whether the use is zootechnical, hence commercial or therapeutic, and so to the benefit of the health and welfare of the animal – were beyond the mandate of the WG. In other words, such an expert group can only provide a scientific opinion, not make policy.

The WG noted that illegal or improper use of growth-promoting substances, in the form of implants and/or in feed, might present an added exposure to humans who consumed meat or meat products from animals so treated. However, they also noted that this would be no different, in terms of risk management, from the illegal or inappropriate use of any other veterinary products and, as such, would be beyond the remit of the WG.

1.2 Biological Effects of Hormones and Endpoints of Health Concern

1.2.1 General Properties of Hormones

Hormones are vital in normal development, maturation and physiological functioning of many vital organs and processes in the body. However, like any other chemicals of natural or synthetic origin, hormones can be toxic to living organisms under certain circumstances. The toxicity may be due to an excess of its normal ("physiological") action. This may be the result of excessive exposure to the substance, for example following absorption of a large dose, or because the physicochemical nature of the substance gives it greater (more "potent") or more prolonged activity of the same type, or because the hormonal action (endocrine effect) occurs at an abnormal time during development or adult life, or is an action on an organism of the inappropriate sex. Hormones, like other chemicals, may also exert direct toxic actions not related to their endocrine ("physiological") effects.

Hormones are very "active" substances, whereby relatively small doses may have profound effects. However, this does not mean that they do not follow typical dose-response relationships, or that all their actions are necessarily permanent in the current or future generations. The relationship between the dose applied to an organism and the effect produced may be linear or it may follow a more complex relationship because it will depend on the extent of absorption of the substance, how it is transported, metabolised and excreted from the body ("pharmacokinetics") and its access across internal barriers to its sites of action. The severity and duration of hormone-induced toxicity will therefore be related to the local concentration at the target site in the body, at least above any possible minimum threshold for an effect and below the concentration at which the mechanism causing the effect becomes saturated and the action reaches its maximum. Many toxic actions are reversible once exposure ceases because normal physiological processes and repair mechanisms return the affected cells and organism to normality. However, irreversible damage may result if there is extensive damage to a tissue.

1.2.2 Health Endpoints of Concern

Endogenously produced sex steroids exert a wide range of biological effects on the body with most tissues/organs affected to a greater or lesser degree (not just the reproductive organs). These effects vary according to age and gender. Therefore, exposure to exogenous steroidogenically active compounds at certain levels has the potential to affect many organs.

Overwhelming evidence suggests that sex steroids exert effects that are dose-dependent and that a threshold dose exists, below which no biological effect will occur. This threshold may vary according to age, gender and tissue/organ.

Production, bioavailability/metabolism and action of endogenous sex steroids are closely controlled. Exogenous exposure to synthetic sex steroids may therefore be compensated for such that no biological effect ensues.

During development, sex steroids play an *organisational* role that involves programming of various tissues/organs in a gender-specific way. These effects are largely irreversible (*e.g.* sexual differentiation of the external genitalia).

Differences in the degree of exposure to sex steroids during development between individuals can occur naturally (*e.g.* higher exposure in twin pregnancies) and this may alter predisposition to future disease such as breast or testicular cancer.

Importantly, for the six hormonal substances considered by the WG, effects of concern would have thresholds, although they may be difficult to define. As noted above, because of "feedback" systems within the body, the introduction into the body of an exogenous source of a sex steroid hormone may be compensated for, so that no biological effect is produced. However, this may not apply to the foetus, postmenopausal women or pre-pubertal children – thus making these groups potentially more vulnerable.

The major areas of concern expressed in the literature and in the SCVPH Opinions were to health effects of hormonal substances in bovine meat and meat products related to **cancer**, **mutagenicity** and **reproductive effects**, in particular **endocrine disruption**. Generally, cancer and mutagenicity are well described, reasonably well understood by most readers and need little general description. However, endocrine disruption has become, in recent years, an area where there has been concern about potential harmful outcomes for a wide range of chemicals hitherto unsuspected of causing such effects. Many of these are less well described and are thus described below.

1.2.3 Endocrine Disruption

Hormones such as androgens and oestrogens play important roles in the day-to-day functioning of the body. Effects are not restricted to the reproductive system but are pervasive, affecting most tissues in the body, including the brain, bone, muscle, liver, fat, cardiovascular and immune systems.[5] During development, androgens and oestrogens also have important organising effects in which the function of certain tissues may be permanently altered. The role of androgens in masculinising the male (reproductive system, genitalia, brain and rest of the body) during foetal life is the most dramatic example of this. It is well established that when there is inappropriate production or inhibition of normal androgen/oestrogen production/action, whether in foetal, childhood or adult life, then important health disorders are likely to occur.[5]

Against this background, it is understandable that there has been widespread concern about the potential health consequences that might result from human exposure to environmental chemicals that possess intrinsic oestrogenic, androgenic or anti-androgenic activity. In addition to xenobiotics, such effects are also attributed to phytoestrogens.[iii] These chemicals have been loosely termed "endocrine disruptors", based on their potential to alter normal hormone action.[6,7]

Whether or not this potential is realised in the body of the recipient is dependent on many factors, one of which is the potency of the chemical in question, *i.e.* what dose of compound will exert a detectable effect? The oestrogenic potency of any ingested compound is determined by a combination of factors. These include:

- absorption, metabolism, entero-hepatic recirculation,
- binding to plasma proteins such as sex hormone-binding globulin (SHBG),[iv] and
- affinity for binding to either oestrogen receptor-α (ERα) or oestrogen receptor-β (ERβ).

Other factors such as rates of breakdown of the compound and local availability in particular tissues may also be important. It is not easy to predict from simple measurements of any one of these parameters what the overall oestrogenic potency of a compound will be in the whole animal. Most studies on the oestrogenic potency of compounds, to which there is human exposure, have utilised *in vitro* cell transfection systems that assess the ability and affinity of the compound to compete for binding to oestrogen receptors, ERα or ERβ. Compounds that have a high affinity for these receptors are likely to be potent oestrogens and may, therefore, exert biological effects on oestrogen target tissues. However, as other factors mentioned above can influence potency, studies involving *in vivo* administration of such compounds provide the most accurate guide as to whether or not they may have target tissue effects. The immature rat uterotrophic assay is the most widely used endpoint of oestrogen bioactivity *in vivo*. Though this provides perhaps the most useable measure of oestrogenic potency *in vivo*, activity

---

[iii] Phytoestrogens are a group of chemicals produced naturally by certain edible plants. The commonest are the isoflavones (*e.g.* genistein, found in soybeans and legumes), coumestans (found in young sprouting legumes), lignans (found in linseed, many cereals, fruits and vegetables) and prenylated flavonoids (found in hops and beer). Human dietary exposure can therefore be substantial, although very variable depending on the composition of the diet. A recent in-depth review of the evidence for both adverse and beneficial effects of dietary phytoestrogens concluded that many of the available data are equivocal and fail to distinguish between effects of the compounds themselves and effects of other dietary constituents.[8] (COT 2003). However, it was also emphasised that the nature and extent of any *potential* effects of phytoestrogens in the diet will depend critically on both the degree of exposure and the age at exposure.

[iv] In humans, sex steroids do not circulate in the bloodstream in a readily bioavailable form. The majority (>95%) are bound to SHBG or other plasma proteins, such as albumin. An equilibrium exists between the amount of protein-bound and free sex steroid in plasma.[9] (Hammond 2002). Although endogenously produced sex steroids bind to SHBG, many synthetic steroidal compounds do not, *e.g.* the potent oestrogens diethylstilboestrol (DES) and ethinyl oestradiol. The possibility that ingested compounds may bind to SHBG and displace already bound endogenous sex steroid, thus making the latter bioavailable, must also be considered. Theoretically, a compound with minimal or no intrinsic oestrogenicity itself could induce oestrogenic effects if it was able to bind with higher affinity than oestradiol to SHBG and thus displace it.

in this assay does not necessarily mean that there would be similar biological activity at other oestrogen target sites, such as in the breast or in bone.[10]

Most endocrine disruptors have only very weak intrinsic hormonal activity when compared with the natural (endogenous) hormones (testosterone, oestradiol) that are made within the body.[5] It is also commonly overlooked that all hormonal systems in the body are tightly controlled and this usually involves a feedback "balancing system" or systems that constantly check the extent of action of the hormone in question and adjust its concentration up or down accordingly.[5] Therefore, in theory, exposure to exogenous endocrine disruptors at a concentration sufficient to cause an effect should be compensated for by altered production of the endogenous hormone in question.

An important exception to this principle is exposure in foetal/perinatal life when hormones are exerting organisational effects and these feedback systems may not be operative.[5] Also in postmenopausal women, no ovarian oestrogen synthesis occurs and the residual oestrogen production, which occurs predominantly in subcutaneous fat, is not subject to significant feedback control. Thus in these circumstances, ingestion of exogenous hormone will lead to additive increments of exposure. This is something that was given particular attention by the WG in their deliberations.

Some compounds may possess no intrinsic hormonal or anti-hormonal activity, yet still be capable of exerting hormonal effects. This can occur if the compound affects the production, bioavailability or metabolism of endogenous (potent) hormones. Such compounds potentially pose a more serious health threat, as alterations in endogenous hormones will cause clinical disorders; there is a growing number of examples of such compounds.[11] Evaluation of potential effects of residues of hormone growth promoters in meat ingested by humans therefore needs to be considered against the background outlined above. In contrast to most endocrine disruptors, several of the growth promoters are intrinsically potent hormones (*e.g.* oestradiol). This means that effects are more likely if there is significant human exposure. However, human exposure is most likely to residues of the growth-promoting agent present in muscle/fat, which will normally be at very low concentrations (pg to $\mu g\, kg^{-1} day^{-1}$). This may rule out possible effects.[12–14] Moreover, as human exposure will be *via* food, the absorption and metabolism of the compound in the gut becomes very important. Oral absorption of oestradiol is good, however the quantity reaching the systemic circulation is greatly reduced by extensive first-pass metabolism in the intestines and liver, and oestradiol is generally considered to be inactive when administered orally.[15]

1.2.4 The Use of Oestradiol in Cattle

Cattle are the main food-producing species in which oestradiol products are used for therapy or growth promotion. In order to put the contribution to the food chain from therapeutic and zootechnical use of oestradiol in context, the endogenous production of oestrogens arising at various stages of the reproductive cycle is described below.

1.2.4.1 The Reproductive Cycle of the Cow

The reproductive cycle of the "average" dairy cow, calving approximately once a year, involves four to five oestrous cycles followed by pregnancy. On average, she spends approximately 75% of the year being pregnant and produces milk for all but the last 40–60 days of pregnancy.

Endogenous production of oestradiol and oestrogens varies throughout the reproduction cycle. In the "cycling" cow there are two or three small peaks of oestradiol during the 21-day oestrous cycle, which accompany waves of follicular development, and one major peak at oestrus. Reported concentrations of oestradiol in plasma and milk vary according to the assay method used but are typically four to five times higher during oestrus than in the remainder of the cycle. During pregnancy oestradiol concentrations in plasma and milk rise dramatically and are typically ten-fold higher than in the cycling animal. Therefore most milk comes from pregnant animals and thus contains higher concentrations of natural endogenous oestrogens.

In addition, a proportion of cows/heifers entering the food chain are pregnant. Meat from these individuals can also contain higher concentrations of oestrogen produced by the foeto-placental unit. The WG noted that when the removal of the ban on the inclusion of meat from cattle over 30 months into the food chain occurs, approximately 25% of cull cows entering the food chain are likely to be pregnant.[16] Meat from these animals will add significantly to the oestrogen concentrations currently entering the food chain from this source. However it should be noted that removal of the restriction would only return the oestrogen load to pre-ban concentrations.

1.2.4.2 Therapeutic Use of Oestradiol in Cattle

The uses and indications for oestradiol salts have been recognised for some time and are clearly defined. Oestradiol benzoate was authorised for the treatment of pyometra and endometritis in cattle. Therapy may also be beneficial to enhance oestrous behaviour in suboestrous or anoestrous animals in the induction of lactation as well as in the dilation of the cervix in cases of abortion. Various oestradiol salts were also luteolytic and are incorporated into oestrous synchronisation devices (PRIDs). Equally, the administration of 1 mg of oestradiol by injection in conjunction with the intravaginal progesterone-releasing device (Cidr) increases synchrony and may enhance the expression of oestrous behaviour. However, since the ban noted above, no oestradiol-based medicines are available in the UK except mescalin for use in dogs.

1.2.4.3 Intra-uterine Infections after Calving

Post-partum endometritis occurs mainly in dairy cows. Various reports estimate that prevalence in these animals ranges between 3 and 8%. This condition occurs during early lactation when discharges need to be eliminated to aid hygienic conditions for milk production.

A proportion of the cases occur in cows that have already experienced a post-calving ovulation and have functioning corpora lutea. A proportion of these will respond to a luteolytic injection by producing endogenous oestrogens and resolving the condition by "self-cure". However, a significant proportion simultaneously experience post-partum anoestrus due to ovarian inactivity. This subgroup is not suitable for luteolytic therapy. For this group there are two alternative strategies. The first uses an intramuscular injection of oestradiol benzoate to mimic the effects of normal ovarian follicular cyclicity. The result is relaxation of the cervix, improved muscle tone and increased supply of leukocytes to the uterus. These induced changes result in evacuation of the uterine contents and elimination of infection. Following the injection, the blood concentration of oestradiol does not rise above the normal physiological range. Indeed, as the remnants of the foeto-placental unit are a source of oestrogen, early evacuation may result in a more rapid fall in milk oestrogen concentrations in these cows. The second accepted therapy involves intra-uterine infusion(s) of an antimicrobial (often an antibiotic) solution. This approach aims to reduce the intra-uterine infection and thus promote a return to normal ovarian cyclicity.

An overall ban on the therapeutic use of oestradiol and/or its esters would thus prevent the former therapy and greater use of antibiotics would be necessary. As these cows are producing milk that may enter the food chain, minimal use of antibiotics is required. Also a proportion of these infections may not respond to antibiotic therapy. Therefore, the result of a ban would be an increased risk of antibiotic resistance and reduced standard of welfare for a proportion of cows.

1.2.4.4 Veterinary Medicinal Products Containing Oestradiol

At the time of the WG's activity in 2002, and before the current ban, there were four veterinary medicinal products containing oestradiol available in the UK as shown below in Table 1.1. These products were formulated to release their oestradiol content in one burst of short duration and are therefore not suitable for growth promotion. One of these products was licensed exclusively for use in the bitch whilst the other three are licensed for zootechnical and therapeutic uses in the cow. Crestar devices were used purely for oestrous synchronisation in dairy and beef heifers and beef cows, and are not used in lactating dairy cows. PRIDs were used both for treatment of suboestrous and anoestrous as well as for oestrous synchronisation in both beef and dairy animals.

1.2.4.5 Food Safety Considerations Following Application of Oestrus Control Products

Crestar was not licensed for use in lactating dairy cattle and, despite the longer half-life of oestradiol valerate, the withdrawal period should ensure that no residues should reach the food chain *via* meat. In contrast PRID and oestradiol

Table 1.1 Example of oestradiol containing veterinary medicinal products

Product	*Active ingredient*	*Conc^n^* $(mg\,ml^{-1})$	*Route of administration*	*Indications*	*Dose*	*Withdrawal*
Mesalin	Oestradiol benzoate	0.2	Subcutaneous or intramuscular	Mesalliance in the bitch	0.01 mg kg^{-1} 3 and 5 days post mating	N/A
Oestradiol benzoate	Oestradiol benzoate	5	Subcutaneous (bitch)	Mesalliance in the bitch	0.3 mg g^{-1} 1–4 days post mating	N/A
			Intramuscular (cow)	Pyometra and endometritis in the cow	3 mg 500kg^{-1}	Milk – 0 d Meat – 15 d
Crestar	Oestradiol valerate	2.5	Intramuscular	Oestrous synchronisation in combination with implant	5 mg	Not for use in milking cows Meat – 14 d post implant removal (23–24 d)
PRID	Oestradiol benzoate	N/A	Intravaginal	Oestrous synchronisation and stimulation of ovarian activity in anovulatory and suboestrous cows	10 mg	Milk – 0 d Meat – 24 hrs

benzoate were licensed for use in both dairy and beef animals. However, as long as withdrawal periods are observed there were considered to be no residue implications associated with these products. The only area of concern noted would be the intramuscular injection site where significant residues may be present if the withdrawal periods were *not* observed. (It is, however, worth noting that therapeutic doses of oestradiol result in pg concentrations that have a half-life of 8 hours and do not exceed normal endogenous blood concentrations.)

The rationale for continued use was that, used for therapeutic or zoo-technical reasons, these products do not cause the concentration of plasma or milk oestradiol to rise outside the physiological range. The use of oestradiol benzoate will cause an elevation in plasma and milk oestradiol concentrations; however these elevated concentrations are still well below those of naturally circulating oestradiol in pregnant animals.

As an illustration, the WG reported that if one considers the total number of treatments with oestradiol benzoate and PRID in 2002 and assumes that they were all delivered to lactating dairy cattle, and take the worse-case scenario, this use elevates oestradiol concentrations to the equivalent of a pregnant animal for 2 days. In 2002 there were approximately 94,500 treatments sold (oestradiol benzoate 29,870 doses; PRID 64,448 doses); assuming these were all used in the 2.25 million dairy cows in the UK, this equates to 0.042 doses per animal. The worse-case impact of this use could therefore be said to be the equivalent of extending pregnancy by 2 hours per cow in the national herd. To put this into context, a 1.2% increase in the proportion of dairy cows in calf to continental beef bulls would result in a similar increase in the duration of pregnancy (by virtue of the longer gestation period of these breeds).

Interestingly, the WG Report noted that the last 30 years have seen an increase in the use of continental sires from virtually zero to approximately 30% of dairy cows services; this management change alone had resulted in an additional 5.25 million days of pregnancy or 2.33 extra days per cow – equivalent to some 2.65 million oestradiol treatments per year.

1.2.4.6 Growth Promotion – Multiple Implantations into Cattle

Implanting hormonal growth promoters was and is currently widespread in the beef cattle industry of many non-EU countries for the better performance in growth and improvement of feed efficiency. In 1999, more than 96% of all US cattle in feedlots were implanted at least once (NAHMS, 2000, cited in 17). These hormonal implants may enhance growth during suckling, growing and finishing stages of production.[18,19] Implant residues entering the food chain as a result of the implants administered during the suckling and growth phases of production will be lower than those arising from implants during the finishing stages. The weight gains are significant. A combination of trenbolone acetate and oestradiol improved average daily gain and feed efficiency during finishing stages by up to 20% and 13.5% respectively.[20] The magnitude of the response to these anabolic implants in the performance of beef cattle varies depending on the type of implants, quantity of growth promoter, duration of exposure, age of animals and combination of implants. Improved performance in steers originating from the dairy herd has also been noted.

In general, anabolic implants have minimal or negative effects on meat quality including lower marbling, high shear force and advanced carcass maturity resulting in lower quality grades. Repeated (five sequential implants) implanting has been claimed to have detrimental effects detectable by consumer taste panel scores.[19] However, consumers failed to detect these differences in meat after 7 and 14 days aging when more moderate (two sequential implants) implant regimes were used.[21] Therefore, there is no organoleptic characteristic by which consumers can be expected to detect meat originating from implanted animals. To date, the WG noted that there was no validated technique to detect and assign the low residual concentrations of oestradiol in the finished edible products to natural sources or to implant residue. This is an area where research was noted to be urgently needed.

As noted above, these implants are no longer allowed in the European Union (EU), which also prohibits the importation of beef and its products derived from hormone-treated cattle.

1.2.4.7 Alternatives to Oestradiol-containing Products

In the absence of oestradiol-containing products, alternatives would need to be employed. For oestrous synchronisation regimes prostaglandin or the progesterone-releasing device (Cidr) could be employed. Alternatives for the treatment of pyometra and endometritis could include the use of prostaglandins for a combination of their direct ecbolic and luteolytic effects. However, it would not be possible to substitute for the current "off-label" use for enhancement of oestrous behaviour.

1.2.4.8 Zootechnical versus Therapeutic Use

The WG was of the view that the growth-promotion activity should be seen separately from the other zootechnical uses and the therapeutic uses of 17β-oestradiol and other hormonally active substances. One strongly expressed view was that, if the current EU position on the ban of 17β-oestradiol for growth-promotion purposes were to be maintained and extended, it would be most unfortunate to lose its use for other zootechnical or therapeutic purposes. There was also agreement that the therapeutic uses of 17β-oestradiol were more important than other zootechnical uses. It was noted that many of the alternatives to 17β-oestradiol would also result in a comparable rise in endogenous oestradiol. As an example, the use of prostaglandins, if used as an alternative, would raise endogenous oestradiol concentrations, so having a similar outcome to the administration of 17β-oestradiol in the first place. For this reason alone, it seemed sensible to continue with the use of 17β-oestradiol. It is well established that prostaglandins can exert both respiratory and reproductive effects following accidental exposure; for this reason it was felt that the operator risks associated with the use of prostaglandin products should not be overlooked.

1.2.4.9 Implications of Removal of Oestradiol-containing Products

Finally, the WG noted that it was important to consider the implications of removal of the use of oestradiol-containing products in food-producing species. Some of the possible implications of the removal of oestradiol products are:

- an increase in the use of prostaglandins, which have health and safety implications for the operator as well as increasing endogenous oestrogens,
- an increase in the use of antibiotics for the treatment of endometritis,
- the development of microbial resistance due to increased use of antibiotics,
- welfare implications through sub-optimal treatment of affected cattle,
- "off-label" use of oestradiol-containing products licensed for use in companion animals is likely to occur,
- unregulated use of oestradiol formulated on an *ad-hoc* basis from chemical suppliers may occur.

1.3 The Scientific Evidence Available to the VPC Working Group

The following sections contain a précis of the scientific evidence used by the WG in its deliberations and presented in their report. It consisted of discussion of the papers and reports emanating from the 17 EC-funded studies as well as other relevant studies identified by members of the WG. It also contains the conclusions of the WG within each of the specific areas.

1.3.1 Exposure to Hormonally Active Substances

1.3.1.1 Analytical Techniques

The SCVPH 2002 Opinion discussed four of the 17 EC-sponsored studies which concerned analytical techniques for the detection of trace hormones in meat and one study which developed screening bioassays for known oestrogenic and androgenic compounds in yeast, trout hepatocytes and human endometrial cancer cells (*Presence of oestrogen in meat*) would have been of particular relevance, but the WG were informed that no publication would be forthcoming.[V] Two studies, both entitled *Analysis of 500 samples for the presence of growth promoters* would appear to represent key research involving new

[V] Confirmed in a letter from Dr Belingieri, 17 July 2003.

methods for the detection of trace hormones in meat, based on GC/MS. However, the report of one comprised only a one-page abstract of a lecture, the text of which includes a number of anecdotes but no new study data. The other study was supported by two publications.[22,23] The derivation of new laboratory methodology was adequately described. But other than the description of steroids in four samples of residue-positive meat and liver, there are no data on samples that reflect the concentrations of the compounds under consideration in a representative set of samples.

On the basis of the results from these last four studies, the SCVPH report concluded, appropriately, that "the low number of samples does not allow a qualified validation of typical characteristics such as sensitivity, specificity, accuracy and reproducibility".

1.3.1.2 Bioassays for Screening

The SCVPH 2002 Opinion discussed one study that developed a screening bioassay to detect known oestrogenic and androgenic compounds in yeast, trout hepatocytes and human endometrial cancer (Ishikawa) cells.[24] The study revealed a highly variable sensitivity between the tests for oestradiol, and a variable differential response in *in vitro* potency tests that may in part be explained by the metabolism of some of the compounds by trout hepatocytes and Ishikawa cells. No data were derived by application of these techniques to meat; if they were to be so applied, exhaustive chromatography to isolate individual steroids would be required in order for the tests to provide useful data. The SCVPH report concluded that, in view of their lack of specificity and sensitivity, the assays performed in recombinant yeast and trout hepatocytes are not justified. The SCVPH Opinion of 2002 on the unsuitability of the yeast assay seemed reasonable to the WG. The WG also noted that this is a profoundly different conclusion from the SCVPH Opinion of 1999, when it was the availability of this highly sensitive (as it was then regarded) new bioassay that led them to consider that previous data on low oestrogen concentrations might be flawed.

1.3.1.3 WG Conclusions and Recommendations

The Working Group concluded that a number of new analytical methods had been developed that might helpfully be applied to the analysis of residues in the meat of cattle, but no substantial data had been presented from their application, nor had they been fully evaluated. These new techniques should be applied to meat in sufficient sample sets to provide reliable estimates of the relevant residues in untreated and implanted animals in the form that they enter the human food chain.

The suitability of three complementary bioassays for screening tissues for oestrogenic and androgenic compounds had not been demonstrated. Unless rigorous chromatographic separation techniques are developed, these bioassays should not be used for assessing residues in meat.

1.3.2 Bioavailability of Hormonally Active Substances

The SCVPH 2002 Opinion discussed two EC-funded studies relating to the bioavailability of hormonally active substances. One study[14] involved the development of a new assay procedure for quantification of oestradiol concentrations in edible tissues and subsequent measurements of oestradiol concentrations in tissue samples from cattle following oestrogen implantation. The new assay included the analysis of lipoidal esters of oestradiol. Validation of the analysis of free oestrogens was complete but was only partial for the analysis of lipoidal esters. Nonetheless, the conclusion that lipoidal esters account for approximately 50% of the total oestradiol concentration in control or single-implanted steers appears sufficiently sound, as is the conclusion that this fraction should be taken into account when assessing the overall intake of oestrogens from treated cattle. Another study[25,26] investigated the metabolism of 17β-oestradiol by bovine hepatocytes and human intestinal and breast cells and tested their oestrogenic properties in the rat uterotrophic bioassay. These studies showed that 17α-oestradiol as well as lipoidal esters of 17β-oestradiol may be formed *in vivo* in animals implanted with 17β-oestradiol as a growth promoter. 17α-oestradiol had only about 10% of the *in vivo* oestrogenic potency of 17β-oestradiol, whereas the lipoidal oestrogens had ten-fold higher potency than 17β-oestradiol when tested *in vivo* in the rat uterotrophic assay.

Based on these above studies, the SCVPH 2002 Opinion concluded that metabolism of 17β-oestradiol in bovine species results in the formation of lipoidal esters, and that these esters are largely disposed of in body fat and may contribute significantly to an additional oestrogen exposure *via* meats. Lipoidal oestrogens may have higher potency in the breast due to their postulated transport *via* the lymphatic circulation and might potentially bioaccumulate in edible fat or meat. However, their oral bioavailability in humans following dietary exposure *via* contaminated meat products is unknown.

According to the WG, these studies on 17β-oestradiol metabolism and evaluation of oestrogenic potency *in vivo* appear to have been well conducted. The demonstration[25] that certain residues may have potency in the uterotrophic assay is suggestive of bioavailability *in vivo* at this particular oestrogen target site. But it remains unclear whether similar actions would occur at other sites and whether any biological or "adverse" effect would result. Since the 1999 SCVPH Opinion, more recent data[14] have shown that in steers implanted with one (normal practice) or with two or four implants inserted simultaneously (misuse), dose-dependent increases in concentrations of lipoidal oestrogens are found in fat, ranging from 30–40 ng kg^{-1} (one implant) up to 100–140 ng kg^{-1} (four implants). Similar or slightly higher concentrations of 17β-oestradiol were detected and much lower concentrations of 17α-oestradiol. In muscle, concentrations of all three compounds were generally <100 ng kg^{-1}, whereas relatively high concentrations of 17α-oestradiol were detected in liver and kidney samples (200–800 ng kg^{-1}). Based on the rat uterotrophic studies reported,[25] no significant effects were detected *in vivo* for any of these three oestrogens at doses of 25 nmol kg^{-1} day^{-1} (~7000 ng kg^{-1}) over a 6-day period.

Assuming similar absorption and metabolism profiles in the human and rat, these findings would suggest that consumption of meat/fat from 17β-oestradiol-implanted cattle is unlikely to provide biologically significant oestrogenic exposure, even if unusually large amounts, from animals bearing four times the recommended number of implants, were eaten regularly. However, this conclusion makes numerous assumptions relating to absorption, metabolism and bioavailability and takes no account of the (theoretical) possibility that lipoidal oestrogens might bioaccumulate over time in fatty tissue, such as in the breast. The Maume *et al.* (2001) paper[14] also considered the concentrations of oestrogens in animals with multiple implants in relation to the maximum human daily dietary intakes. For adults their estimates indicate an intake of $<5\%$ of ADI from a standard 500 g meat intake, but for pre-pubertal boys the ADI might be exceeded.

The WG felt that the SCVPH conclusion on the formation of lipoidal esters based on *in vitro* oestrogenic activity expressed in T47 D breast cancer cells$_{26}$ (Hoogenboom *et al.*, 2001) was reasonable, although it is not clear to what extent hydrolysis of lipoidal esters occurred before binding to ER in T47 D cells, and thus did not reflect a direct effect of the esters themselves. The degree to which any such hydrolysis would occur in humans is unknown.

1.3.2.1 WG Conclusions and Recommendations

Theoretically, if considerable amounts of 17β-oestradiol or lipoidal oestrogens are present as residues or contaminants in hormone-treated meat samples, they could exert significant effects on important oestrogen target tissues such as the breast. From the information available it appears that such exposures are unlikely to occur, even in situations in which misuse (*i.e.* over-implantation) of implants has taken place. However, the WG added that this conclusion should be re-assessed when, and if, new data become available to show that bioaccumulation of lipoidal oestrogens in fatty tissue occurs *in vivo* after oral administration. The data from Maume *et al.* (2001)[14] in relation to pre-pubertal boys should be confirmed by others and if confirmed may be a cause for concern in this group.

1.3.3 Cancer Risks of Oestrogenic Substances

1.3.3.1 Breast Cancer Risk

Over the last few years there have been a number of publications that have had a substantial impact on our thinking on the effects of endogenous and exogenous oestrogens on the incidence of breast cancer. The data are directly relevant only to postmenopausal women.[vi]

The Endogenous Hormones Breast Cancer Collaborative Group (2002) collated and analysed data from the nine published studies on the relationship between the plasma concentration of steroid hormones in postmenopausal women and the risk of subsequent development of breast cancer. Statistically significant relationships were found for several hormones. The strongest associations were for 17β-oestradiol (stronger still when only the protein-free fraction was considered) and testosterone. The relationship with testosterone was markedly weakened after adjustment for 17β-oestradiol; this is consistent with the relationship being determined by conversion of testosterone to 17β-oestradiol by the action of the aromatase enzyme. The Collaborative Group estimated that the relative risk for breast cancer was 1.25 for each doubling of plasma 17β-oestradiol concentrations. It is, however, near certain that this underestimates the true risk,[vii] since only a single blood sample from each subject was analysed in each of the studies. Additionally, the studies used a wide variety of analytical techniques, some of which were inaccurate and/or inappropriate for use in postmenopausal women.

The WG noted that recent publications from two very large studies have confirmed that use of hormone replacement therapy (HRT) by postmenopausal women for several years significantly increases their risk of breast cancer.[28,29] The Women's Health Initiative (WHI) conducted a randomised, placebo-controlled trial of combination HRT (oestrogen plus progestin) *versus* no HRT in 16,608 North American women and found that breast cancer incidence was increased with a hazard ratio of 1.24.[29] Notably, this study also reported that the breast cancers presented at a significantly later stage in the HRT users. The Million Women Study (MWS) recorded HRT usage in around one million postmenopausal women in the UK.[28] Consistent with the WHI study, a higher incidence of breast cancer was found in combination HRT users. Importantly, in the context of the possible impact of ingested exogenous oestrogenic residues, MWS also reported a higher incidence in women taking oestrogen-only HRT (relative risk of 1.30 *versus* never users). These two studies confirmed that incidence of breast cancer in postmenopausal women is enhanced by the regular ingestion of oestrogens, mainly in the form of oral conjugated equine oestrogens, in quantities sufficient to reduce menopausal symptoms and preserve bone integrity. However, it is not possible from these studies to ascertain a concentration of 17β-oestradiol that does not enhance the risk of breast cancer (*i.e.* an NOAEL cannot be established).

Yen *et al* (1975)[30] had described the effects of ingested oestradiol on plasma oestradiol, oestrone and gonadotrophin concentrations in postmenopausal

---

[vi] In postmenopausal women, no ovarian oestrogen synthesis occurs and the residual oestrogen production occurs predominantly in sub-cutaneous fat and is not subject to significant feedback control; in these circumstances, ingestion of exogenous hormone leads to additive increments of exposure.

[vii] It has been estimated by the evaluation of multiple samples that the true relative-risk from plasma oestradiol is double that estimated by single-sample studies.[27] (Hankinson *et al*, 1995). Thus, the relative risk of breast cancer from a doubling of oestradiol concentrations is likely to be approximately 1.50.

women. Their data showed that 2 mg micronised oestradiol led to a maximum plasma concentration of $110\,pg\,ml^{-1}$ 5 hours after ingestion and this was a 437% increase *i.e.* from a starting concentration of $20\,pg\,ml^{-1}$, falling to baseline by 24 hours. Thus, the increment in oestradiol from ingesting 2 mg was $90\,pg\,ml^{-1} = 320\,pmol\,l^{-1}$. Over 24 hours the mean increment would be no more than $150\,pmol\,l^{-1}$.

The highest concentration of oestradiol detected in meat was $56\,ng\,kg^{-1}$ in kidney.[31] For a postmenopausal woman eating a kilogram of such kidney from a treated animal, the theoretical increment would therefore be $0.004\,pmol\,l^{-1}$ (based on the finding that the 97.5th percentile of consumers eat 40 g of kidney per day). This increment is approximately three orders of magnitude below the most sensitive assays available and below any concerns related to breast cancer risk. Assuming that all of the oestradiol in meat is bioavailable and unaffected by food preparation, this would be expected to lead to mean concentrations increasing from approximately $40\,pmol\,l^{-1}$ to $40.004\,pmol\,l^{-1}$, an increase of only 0.01%.

1.3.3.2 Reduced Breast Cancer Risk in Future

The WG noted that in the UK in 2002, there were over 100,000 women receiving tamoxifen for treatment of breast cancer, of whom about 75% were postmenopausal. Modern aromatase inhibitors (*e.g.* anastrozole, letrozole, exemestane) have shown themselves to be superior in efficacy to tamoxifen,[32] and it was widely expected that in the next few years this population will instead be treated with aromatase inhibitors. The efficacy of these compounds is determined by their suppression of plasma oestradiol concentrations from $25\,pmol\,l^{-1}$ to below the detection limit of available assays ($<3\,pmol\,l^{-1}$). However, their efficacy could be compromised by the ingestion of oestradiol in doses that achieved increments of plasma oestradiol in single figures of $pmol\,l^{-1}$.

1.3.3.3 In utero Exposure and Breast Cancer Risks

Evidence to support a role of intra-uterine factors such as 17β-oestradiol concentrations and development of breast cancer in the female is well established.[33,34] One of the approaches used to explore the potential involvement of hormones in affecting predisposition to cancer is to compare twin *versus* singleton pregnancies, as oestrogen concentrations in twin pregnancies are invariably higher than in singleton pregnancies.[35] The SCVPH report discussed one EC-commissioned study based on the Swedish Twin Registry that sought to evaluate whether risk of breast cancer was higher in twins.[36]

1.3.3.4 Oestrogen and the Human Gut

Ingestion of meat from animals treated with hormonally active substances is likely to result in highest levels of exposure in the gut. Therefore, potential effects of oestrogenic and/or androgenic compounds on the gastrointestinal tract need to be considered. There are clear gender-related differences in gastric acid production (40% higher in males[37]) and in the incidence of gastro-duodenal ulcers (higher in men[38]), Crohn's disease (higher in females[39]) and colorectal cancer (higher in males than in pre-menopausal women[40]).

There is reasonably convincing evidence that the gender difference in gastric acid production and colorectal cancer stems from differences in oestrogen production/action in males *versus* females, as oestrogen treatment reduces gastric acid production,[41] and oestrogen exposure, whether endogenous or *via* hormone-replacement therapy, is protective against colorectal cancer.[40] The precise involvement of oestrogens in progression of Crohn's disease is less clear.[42] Based on these observations, the WG felt that, on balance, it would be concluded that any additional exposure of the gut to oestrogenic compounds present in meat from growth-promoted animals would have a health-beneficial, rather than -detrimental, effect.

1.3.3.5 WG Conclusions

The WG noted that recent studies have confirmed hormone replacement therapy increases the risk of breast cancer in postmenopausal women. However, it also noted that the maximum increase in oestradiol concentrations which might occur following consumption of oestradiol-treated meat by a postmenopausal woman is most likely to be below any concerns related to cancer risk. Oestrogen therapy appears to be protective against colorectal cancer and, therefore, arguably, any additional exposure following ingestion of oestradiol-treated meat would, if anything, have a health-beneficial effect for this commonly occurring cancer.

1.3.4 Altered Gene Expression by Oestrogenic Substances

The SCVPH 2002 Opinion discussed one EC-funded study that measured changes in gene expression of a number of hormone sensitive genes in a breast cancer (MCF7) cell line.[43] The study found the expression of the different hormone responsive genes varied for the different oestrogens (zeranol and five related compounds, 17 β-oestradiol and three other oestrogenic substances). The SCVPH's only comment on this study was to note the down-regulation of GSTμ3, a Phase II enzyme that is involved in protection against DNA damage by free oxygen radicals.

The results showed that zeranol was of similar potency to 17β-oestradiol in this test system, although there were gene-specific differences in the extent of expression following treatment with the oestrogenic substances. Zeranol was much more potent than the naturally occurring fungal contaminant, zearalenone. It was also noted that zeranol, as a mycotoxin, may arise from fungal growth in cereals. Although interconversion of these substances can occur, this occurs at a low rate, suggesting zeranol may pose a greater hazard than the widely occurring zearalenone. Zeranol was the most potent inhibitor of the expression of MRG1/p35srj, which is involved with a nuclear transcription activation factor.

The WG noted that a number of authors have demonstrated that zeranol shows no or only limited binding to cellular binding proteins in contrast to 17β-oestradiol. This indicates that zeranol may be more potent in hormonal activity than 17β-oestradiol, due to higher bioavailability.[44–47] Currently, its bioavailability by the oral route following consumption of meat products containing zeranol is not known. However, the higher oestrogenic potential seen in this *in vitro* test system is not consistent with reports of lower oestrogenic potential in a variety of *in vivo* assays *e.g.* vaginal cornification, uterotrophic assay and depression of serum gonadotrophin concentrations in castrated monkeys (reviewed in Lindsay, 1985). Furthermore, zearalenone and zeranol were shown to have similar physiological effects in a variety of *in vivo* assays. This suggests that this gene expression test system may not be a good indicator of *in vivo* oestrogenic potency.

Leffers *et al.* (2001)[43] also showed 17β-oestradiol down-regulated GST μ3 at extremely low concentrations and suggested that this response might be a result of the altered redox status within the cell, rather than due to regulation by the oestrogen receptor. Together with the down-regulation of other phase II genes, they suggested that this could reduce protection against DNA damage and that changes in the relative balance of gene expression of Phase I and Phase II metabolism may be important in the proposed production of genotoxic catechol metabolites of 17β-oestradiol.

1.3.4.1 WG Conclusions and Recommendations

The evidence that oestradiol gives rise to genotoxic metabolites is considered further later. The low binding activity of zeranol and its ability to alter gene expression of important hormone responsive genes makes it important to determine the bioavailability and biological significance of zeranol residues in meat.[48] Initially this would require studies of serum concentrations following consumption of meat from zeranol-treated animals.

1.3.5 Genotoxic and Mutagenic Effects of Oestrogenic Substances

In its 1999 Opinion, the SCVPH concluded there was sufficient evidence that 17β-oestradiol was genotoxic. This Opinion was based on positive responses in a variety of *in vitro* indicator assays. The VPC Sub-Group report (1999) pointed out that standard mutagenicity tests on 17β-oestradiol (bacterial mutation, mammalian gene mutation, *in vitro* micronuclei, the bone marrow micronuclei test and germ cell cytogenetic assay) were all negative. Furthermore, the studies on which the SCVPH based the opinion were all non-standard studies (methotrexate resistance, microsatellite formation), or were unconvincing due to the absence of a dose-response. The SCVPH concluded, however, that there was evidence for induction of oxidative damage, DNA adducts and aneuploidy.

1.3.5.1 17β-Oestradiol

The 2002 SCVPH Opinion states there is now conclusive evidence that 17β-oestradiol is genotoxic since it induces mutations in mammalian cells, oestradiol metabolites induce mutations in mouse skin *in vivo* and catechol oestrogen quinones form DNA adducts in cultured cells and mouse skin. This SCVPH Opinion is based on a study commissioned by the EC[25,26] and other published papers. The study by Hoogenboom and colleagues showed that 17β-oestradiol and several of its metabolites were negative in a series of apparently well-conducted bacterial mutagenicity assays using a variety of strains and metabolic activation conditions employed in order to improve the potential sensitivity of the test. Furthermore, they also reported negative responses in the Comet assay using human intestinal cells (CaCo-2). The other published papers considered by the SCVPH are discussed below.

A number of recent papers strengthen the evidence that 17β-oestradiol can be activated to produce genotoxic metabolites by its conversion into catechol oestrogens which may be oxidised to form semiquinones and quinones (*e.g.* [49–51]). These quinones can form DNA adducts leading to depurination. The metabolites may also generate potentially mutagenic oxygen radicals by redox cycling. Inactivation *via* O-methylation, glucuronidation or sulfation also occurs.

The mutagenic potential of oligodeoxyribonucleotides adducted with hydroxyoestrogen moieties was studied.[52] A series of synthetic oligonucleotides was produced, each containing a modified nucleotide. These were used to create vectors, which were then used to transfect COS-7 monkey kidney cells. They were shown to induce G to T transversions in this model system. This study demonstrates that oestrogen metabolite adducts introduced into naked DNA are pre-mutagenic. The system bypasses normal cellular controls (activation/inactivation pathways and DNA repair) of the intact animal.

A metabolite of 17β-oestradiol, 2-methoxyestradiol (2-MeE2), was claimed to induce transformation in the SHE assay, chromosome aberrations and mutations at the HPRT and Na^+/K^+ ATPase loci.[53] The study is poorly reported. 2-MeE2 induced mutations at only one of the doses tested, which was a mid-point dose in the case of the Na^+/K^+ ATPase assay. Furthermore, it is not clear whether cytotoxicity has been assessed appropriately and the assays lacked statistical power due to the low control frequency and low numbers of cells analysed. At best the evidence is marginal. The induction of chromosome aberrations is even less scientifically convincing. Concurrent cytotoxicity data are not given, but assuming the concentrations are similar to those measured in the mutation assay, then an increase in aberrations was only seen at toxicity concentrations in excess of the internationally acceptable limits. Nearly all the damage was due to "chromosome pulverisation", an effect attributed by the authors to asynchronous division within multinucleate cells, and not therefore due to clastogenicity. Aneuploidy and polyploidy were also induced. While there is some evidence that the metabolite was able to induce cell transformation, the significance of this finding is less clear as the SHE assay detects both genotoxic and non-genotoxic substances. A wider range of metabolites was tested in the same systems.[54] Despite significant methodological and reporting inadequacies, the WG agreed that there does appear to be some evidence that some of the metabolites can induce chromosome aberrations, mutations and cell transformation in SHE cells. 4-hydroxyestrone and 2-methoxyestrone, but not 17β-oestradiol, oestrone, 2-hydroxyoestradiol or 4-hydroxyoestradiol, induced mutations at the HPRT locus. Estrone, 4 hydroxyestrone and 4-hydroxyoestradiol, but not oestradiol or the other hydroxyl metabolites tested, gave some evidence of weak induction of chromosome aberrations.

The evidence that 17β-oestradiol is a point mutagen is derived from the publication of Kong *et al* (2000),[55] based on the induction of mutations at the HPRT loci in V79 cells. However, this study is not acceptable by generally recognised standards of quality or accuracy. In the study report there is insufficient information to evaluate whether an adequate number of cells was treated, there is no dose-response (significant increases were seen in cultures treated with 10^{-11}, 10^{-10}, 10^{-7} and 10^{-6} M 17β-oestradiol but not with 10^{-8} and 10^{-9} M) and there is no evidence that the protocol ensured the independence of the individual mutant colonies picked for assessing the mutation spectra. The mutation induction data appears to be based on separate experiments, combined into a single results table, making it impossible to determine the data obtained in each separate experiment and thus to see how the reported increases relate to control (or spontaneous) values. Some of the DNA base changes in the "induced" mutants are incorrectly assigned. A key observation is the occurrence of two "hotspots" of mutation. These specific changes are rarely found and furthermore may have arisen due to failure to ensure the independence of mutants selected.[viii] Certainly, the postulated mechanism of action of oestradiol due to free-radical induced DNA damage would not be predicted to produce such a unique profile of DNA base changes. Our current understanding of spontaneous mutation indicates that a major fraction of mutations originates from oxidative damage. Thus, if the proposed genotoxicity of oestradiol is due to oxidative damage, then one might predict a mutant profile similar to those produced spontaneously. Therefore, this study cannot be considered sufficient evidence of mutation potential and the claim that 17β-oestradiol does not act *via* a receptor because mutation is not reduced in the presence of an anti-oestrogen is also not substantiated.

The study of Chakravarti *et al.* (2001)[56] is cited as evidence that oestradiol-3,4-quinone induces mutations in mouse skin *in vivo*. Dorsal skin of SENCAR mice was treated with this metabolite and the mice were sacrificed after one hour to measure DNA adducts and at intervals thereafter for measuring

[viii] Mammalian Gene Mutation Database, available at: http://lisntweb.swan.ac.uk/cmgt/index.htm, accessed 29 March 2009.

mutations in the H-ras gene. The dose used was 200 nmol oestradiol-3,4-quinone; the treated surface area was not defined. The mutations induced were sequenced. N3-adenine adducts (rapidly depurinating) and N7 guanine adducts (slowly depurinating) were seen. It appears that those arising at N3, but not at N7 guanine, gave rise to mutations. Whilst this study provides evidence that a metabolite of 17β-oestradiol can give rise to a genotoxic effect *in vivo*, the mutation frequencies obtained (2.2×10^{-5} mutations per base pair) are extremely high and there is no concurrent measure of toxicity. The relevance of this dose level to concentrations to which humans are exposed from eating meat from treated animals would require further investigation.

In a study not considered in the SCVPH Opinion, Yared and colleagues reported on the genotoxic effects of oestrogens in breast cells using the micronucleus and Comet assays.[57] 17β-Oestradiol, oestrone and oestriol were tested for their ability to induce micronuclei in an assay using cytocholasin B and DNA damage detected by the Comet assay in a human mammary cell line (MCF-7) and primary human mammary epithelial cells, both of which have the oestrogen receptor. Oestradiol induced an increase of micronuclei at 10^{-9} M. Higher concentrations also showed an increase above controls, but in an inverse dose-response. Oestrone induced a dose-related response in micronuclei. No increase was observed for oestriol. A dose-related induction of proliferation was also observed for all compounds. Positive responses in the Comet assay were seen for β-oestradiol and oestrone and to a lesser extent for oestriol in both cell types.

1.3.5.2 Testosterone and Progesterone

Testosterone had previously been reported to be negative in the L5178Y gene mutation assay and in *in vivo* somatic and germ cell assays for chromosome aberrations.[58] No data were available on progesterone. The SCVPH considered further the JECFA/WHO evaluation of these hormones (WHO Food Additives Series 43; WHO, 2000[2]) and considered there is no evidence that progesterone or testosterone have genotoxic potential. No other publications have been published to add to this.

1.3.5.3 Zeranol and Trenbolone

The 1999 SCVPH Opinion concluded that trenbolone was not genotoxic on the weight of evidence from numerous studies. Isolated positive responses were reported for micronucleus induction and cell transformation in SHE cells but not in C3H10T1/2 cells. There were, however, no standard assay results available on zeranol. On the basis of further work commissioned by the EC,[59] the SCVPH 2002 Opinion concluded that these substances exhibit only very weak effects.

The mutagenicity of these substances was also investigated.[59] β-Trenbolone was negative in a cell mutation assay (V79/hprt) and at the lacI loci in *E. coli*.

Marginal positive results were claimed for micronuclei induction in V79 cells and for DNA adducts in hepatocytes. Zeranol did not induce DNA adducts in rat hepatocytes, mutations at the lacI locus in *E. coli* or mutations in mammalian cells (V79/hprt). A borderline response was seen for induction of micronuclei *in vitro*. The positive micronucleus responses for both compounds were only obtained at near-cytotoxic concentrations. The authors conclude that further work is required to evaluate the genotoxicity of these substances and their metabolites, and that non-standard systems may be required to detect weak effects.

The WG considered there to be a number of methodological flaws with Metzler and Pfeiffer's gene mutation study – insufficient cells were treated and maintained through the expression period and assessed for mutations; a single dose only was assessed and the cytotoxicity values are not presented. Similarly the micronucleus results were obtained at a single concentration only and the measures of cytotoxicity were not presented, although it would appear that near cytotoxic concentrations were used. The method did not use the cytocholasin B method and its sensitivity was consequently affected by the inhibition of cell proliferation by trenbolone.

A further study considered by the SCVPH involved the interaction of hormonal substances and their metabolites with sex hormone-binding globulin (SHBG) or the analogous sex hormone-binding protein (SBP) from trout plasma. Zeranol and its metabolites were included in this study, and were found to have low binding affinity to these proteins; this would result in high bioavailability when present in plasma, but also fairly rapid metabolism. This unpublished study is discussed further.

1.3.5.4 Melengestrol Acetate

At the time of the SCVPH 1999 Opinion, the information available on melengestrol acetate (MGA) was sparse. In its 2002 Opinion, the SCVPH considered the recent JECFA evaluation (WHO Food Additives Series 45; WHO, 2000[2]), but noted that most of the references were to unpublished reports. An EC-funded study addressed this issue[59] and showed that MGA was negative in a cell mutation assay (V79/hprt), in a micronucleus test in V79 cells and in a gene mutation assay for LacI mutations in *E. coli*.[59] SCVPH (2002)[4] concluded from this study that MGA showed only weak effects. However, the WG considered the published study provided insufficient information for evaluation and thus no conclusion can be made on the mutagenicity of MGA.

1.3.5.5 WG Conclusions and Recommendations

1.3.5.5.1 17β-Oestradiol. Most of the "new" information referred to in the 2002 SCVPH Opinion has been generated using non-standard methods that produce information of questionable relevance to effects that may occur in the intact animal. A number of the studies discussed in the report are of poor quality. However, there is now additional evidence that metabolites of 17β-oestradiol can form DNA adducts *in vitro*[49,60] and *in vivo*.[49] While the catechol metabolites of 17β-oestradiol induce DNA adducts in SHE cells, 17β-oestradiol itself does not do so.[60] There is some evidence for the induction by oestradiol of DNA damage (single strand breaks) and micronuclei formation in cells with the oestrogen receptor.[57] It is not known whether the micronuclei are due to clastogenicity or to aneuploidy.

Evidence for the induction of mutations by 17β-oestradiol itself has only been obtained in non-standard assays, including those without normal cellular controls (*e.g.* [52]). There is still no evidence that 17β-oestradiol itself is a gene mutagen. The key study of Kong *et al.* (2000),[55] purporting to show that 17β-oestradiol is a gene mutagen, suffers methodological and interpretation flaws. The mutagenicity seen in the *in vivo* study involving skin painting[56] may have been associated with extreme doses. There is, however, some evidence for a clastogenic potential for 17β-oestradiol (reviewed by [49]), although some studies have failed to differentiate between aneuploidy and structural damage. There is further evidence that oestradiol is an aneugen and an inducer of other genotoxic effects (*e.g.* DNA amplification, microsatellite formation). The significance of these latter endpoints for hazard and risk evaluation is still not clear.

Overall, the weight of evidence from many genotoxicity studies, both standard and non-standard, indicates there may be a genotoxic potential for metabolites of 17β-oestradiol, but this direct evidence is by no means substantial. However, there is further indirect evidence for genotoxicity. A plausible hypothesis has been advanced[49,51] that 17β-oestradiol is carcinogenic in humans and animal models by a combination of effects on cell proliferation and by genotoxicity. The hypothesis is primarily based on reasonable evidence that 17β-oestradiol is not carcinogenic solely due to epigenetic phenomena such as induction of cell proliferation.

Although there is evidence that oestrogen metabolites may be directly genotoxic *in vitro*, *in vivo* their formation is affected by opposing activation and inactivation metabolic pathways, the presence of anti-oxidants and DNA repair capacity and thus it is likely this genotoxicity will have a thresholded response. The importance of anti-oxidant defence systems is demonstrated by the reduction in transformation and formation of DNA adducts by oestrogen metabolites in the presence of ascorbic acid.[60]

There were no standard tests conducted *in vivo*, even on 17β-oestradiol metabolites, which indicated a mutagenic potential for 17β-oestradiol *in vivo*. Since DNA repair pathways, anti-oxidant defence and Phase II inactivation pathways can be overwhelmed at high doses, it is necessary to obtain evidence of genotoxicity in well-conducted assays, employing realistic doses.

It is important to determine whether the 17β-oestradiol metabolites can be produced *in vivo* in sufficient quantities to result in genotoxicity. Thus, it is recommended well-conducted *in vivo* studies are performed to determine whether 17β-oestradiol is able to induce genotoxic damage *in vivo* under realistic exposure conditions.

1.3.5.5.2 Testosterone and Progesterone. On the basis of the limited information available to the WG, there was not further evidence of genotoxicity of these substances and no recommendation for further work.

1.3.5.5.3 Zeranol and Trenbolone. The WG concluded that there were insufficient data to indicate zeranol or trenbolone are genotoxic. This conclusion is the same as that reached in the (previous) 1999 SCVPH Opinion. Further studies would be required to evaluate this fully.

1.3.5.5.4 Melengestrol Acetate. The WG considered there to be insufficient data available to evaluate the genotoxicity of MGA.

1.3.6 Developmental and Reproductive Effects of Hormonally Active Substances

The potential for chemicals with intrinsic endocrine activity to perturb development and function of the reproductive system, especially in the male, has been a driving force for concern about the issue of environmental endocrine-active chemicals. Results of studies in experimental laboratory animals have been equivocal, and there are as yet no data to show that such effects occur in humans due to any environmental chemical exposure.

1.3.6.1 Recent Data

The 2002 SCVPH opinion considered three EC-funded studies that addressed reproductive and developmental sequelae of exposure to hormonally active compounds. One was an animal study, involving gestational and lactational exposure of rabbits to zeranol, trenbolone acetate (TBA) and MGA.[61] The other two were human studies: a retrospective case-control follow-up of young men and women suspected of having been exposed to meat from hormone-treated animals when they were children in 1977[62] and a study based on data from the Swedish Twin Registry, looking at breast cancer risks in twins.[36]

Studies to directly assess the effects of zeranol, TBA and MGA on the development of the testis and reproductive system in rabbits were investigated. These studies involved gestational and lactational exposure to these compounds at moderate or high doses, ranging from $0.25\,mg\,kg^{-1}month^{-1}$ by implant for zeranol to $0.5\,mg\,kg^{-1}day^{-1}$ for MGA. Exposure to TBA or MGA was also investigated during adulthood. The SCVPH quote the authors' conclusions as indicating "that prenatal exposure to low doses of MGA, TBA or zeranol may affect the function of the reproductive tract in rabbits, although the effects are not as severe as those observed after exposure to the high doses. The effects are most pronounced if the exposure occurred early in life. All three compounds readily cross the placental barrier and accumulate to a variable degree in fetal tissues. The effects of zeranol and TBA are more severe than the effects of MGA in animals exposed during development; however, MGA has marked effects on spermatogenesis when administered in adults". As only a superficial description of the results of this study was given in the SCVPH 2002 opinion, the WG found it difficult to draw any conclusions with confidence. The only mention of a *significant* change was an increase in concentrations of oestrone after exposure to either MGA or to zeranol during early adolescence, a change unlikely to be of biological significance.

The final report on the above study provided a more conclusive view on this investigation. Pilot studies used relatively high doses of the test compounds and this led to various problems that resulted in use of lower degrees of exposure for the main (reported) study. In the pilot study, treatment with zeranol (dose unspecified, but an implant that delivered $>0.25\,mg\,kg^{-1}month^{-1}$) resulted in major reproductive abnormalities, including cryptorchidism and gross suppression of spermatogenesis. However, no details of this pilot study are given (other than a description of testicular histology) and the use of a lower dose for the main study suggests that it was considered that the pilot study was compromised in some way. Similarly, prenatal exposure to TBA in the pilot studies was confounded by major perinatal mortality of the offspring and only one dam gave birth to offspring that survived after exposure to the lowest dose of TBA ($0.5\,mg\,kg^{-1}week^{-1}$).

In the main study, only minimal effects were observed in animals exposed to doses of the three test compounds during different life phases (for zeranol, a monthly subcutaneous implant of $0.25\,mg\,kg^{-1}$ to the dam; for TBA a weekly subcutaneous injection of $0.5\,mg\,kg^{-1}$; for MGA $0.5\,mg\,kg^{-1}$ orally daily). No consistent significant treatment-related effects were observed, though four cases of unilateral cryptorchidism were observed, two after adolescent exposure to TBA and one after adolescent exposure to zeranol; one animal with unilateral cryptorchidism was observed after gestational exposure to TBA. No other gross changes of the reproductive system were observed. Abnormal spermatogenesis, as evaluated by a non-standard, but published, method, was evident in the cryptorchid testes as expected. But there was no evidence for abnormal changes in scrotal testes (though this is based on deductions from the limited tabulated data provided), with the possible exception of animals exposed in adulthood to MGA. The latter (small) effect most likely occurs as a consequence of the progestational activity of MGA. Even assuming that MGA does have effects on spermatogenesis when administered to adult rabbits, the dose used ($0.5\,mg\,kg^{-1}day^{-1}$) is presumed to have no relevance to humans exposed *via* residues in meat, unless there is ingestion of part of an implant.

Based on cellular morphology, the report refers to the abnormal persistence of "single" "foetal-like" germ cells in the testes of treated animals, although it is not specified in which treatment groups these cells were noted. Such cells are of interest because testicular (germ cell) cancer in humans probably arises from transformed foetal germ cells that have persisted in the testis since foetal life. However, the study was unable to confirm the possible foetal nature of these cells using a battery of specific markers as none of the available antibodies worked on rabbit tissues. Evidence for effects of TBA and zeranol on gonocyte development in the foetal testis was provided by increased numbers of these cells being immunopositive for PG-2, but as the role of this antigen is unknown, it is not possible to evaluate the significance of this observation (which was based on only two animals per group).

Sporadic changes in reproductive hormone concentrations were reported at certain ages in certain treatment groups, but no consistent, easily interpretable pattern was observed; no evidence of major dysfunction of the testis or of the hypothalamic-pituitary-testicular axis emerged from this study. Similarly, no evidence for any change in semen quality was found in any of the treatment groups. From limited studies on maternal and foetal samples from control and treated animals, it was shown that residues of MGA and zeranol and metabolites of TBA were clearly identifiable in various tissues of relevant treated animals, but were not detectable in controls.

This study experienced confounding problems due to "side-effects" of the administered compounds during pregnancy. This is not unusual as pregnancy is susceptible to hormonal disruption, as it is a hormone-dependent process.[63] This may indicate that the doses of the test compounds being used were too high, although this was not a specified conclusion of the report. The use of generally lower doses for the main study largely avoided these confounding problems and provided only sketchy evidence for any significant impact on reproductive development and function as a result of *in utero* or postnatal exposures to MGA, TBA or zeranol. Perhaps the only lingering concern was the sporadic occurrence of cryptorchidism, which was confined to treated animals, though this was restricted to animals exposed during adolescence; this may indicate either that the cryptorchidism was treatment-unrelated (*i.e.* the problem was present prior to initiation of treatment, as cryptorchidism is not uncommon in rabbits) or that the final stage of inguinal testicular descent had been compromised. The latter is well established to be an androgen-dependent process, but the *very limited* data available for testosterone concentrations show no indications of suppression.

Other than the occurrence of cryptorchidism, none of the other findings in treated animals were suggestive of consistent, abnormal changes in development or function of the male reproductive system. Moreover, they occurred in animals in which exposure to the growth-promoting hormones was far in excess of that likely to occur in humans as the result of ingestion of meat/fat from growth promoter-treated cattle. This provides reassurance that adverse effects on the developing human male reproductive system are unlikely to occur.

The SCVPH 2002 Opinion concluded that *in utero* or pre- and peri-pubertal exposure to hormones (including animal evidence on synthetic products) may affect pubertal development and that epidemiological studies with opposite-sexed twins indicated prenatal exposure to hormones may be linked to adult cancer risk. These conclusions derived from two EC-funded studies.

The study of Chiumello *et al.* (2001)[62] was thought by the WG extremely difficult to interpret and has several shortcomings. It followed an outbreak of gynaecomastia in school children in Italy in 1977, when it was ***presumed*** that accidental exposure to an oestrogenic compound of some sort was involved. The source and nature of the compound were never identified. In this situation the presumed exposure mimics what would happen normally during natural puberty when endogenous oestrogen concentrations would rise and stimulate breast development (in the female). If such exposure were continued over a period of time, effects on final height and other parameters might occur that could have significant impact for the individual. Remarkably, in the follow-up study height was not measured (or is not listed on the questionnaire or in the final report). Instead the focus was solely on reproductive issues and only minor changes were found. The most significant finding was an increased incidence of small (atrophic) testes in men who had attended the affected school in 1977. However, even this finding is suspect. First, it is well established that individuals who believe they may have a reproductive problem are more likely to volunteer/participate in studies that involve clinical examinations and blood tests related to reproduction (they get a free check-up); evidence of this was apparent from the report. Second, it is completely unknown whether or not the boys with atrophic testes were "exposed" to the contaminated meat. Third, as this contamination was not proven, nor the nature of any hormonal contaminant identified, it is not possible to draw conclusions from this study.

Even if it accepted that the children in question had been exposed to an oestrogen such as DES, used for meat growth promotion, the WG felt that there was no evidence to suggest that such outbreaks are other than very isolated and rare phenomena. This suggests that exposure of children to oestrogenic compounds in meat is not sufficient to induce precocious breast development to the point where it is clinically significant. The most frequent occurrence of precocious puberty in girls arises in individuals who have been adopted at an early age from a developing country and then reared in a Western country.[64,65] What underlies the extraordinarily high incidence (20–25%) of precocious puberty in such individuals is still unclear but may involve precocious activation of the hypothalamic-pituitary axis.

The study of Kaijser *et al.* (2001)[36] was based on the Swedish Twin Registry and sought to evaluate whether subsequent risk of breast cancer was higher in twins. This study showed that with increasing female birthweight, the risk of developing breast cancer in pre- or post-menopausal life was increased stepwise. Though comparison of twins *versus* singletons can reveal a relationship between oestrogen concentrations in pregnancy and the risk of reproductive cancer in the offspring,[33,34] there are several difficulties in making such associations. First it is unclear what the relationship is between oestrogen concentrations measured in blood of the pregnant mother and those in the foetus, in particular the concentrations in oestrogen target tissues. Second, in twin pregnancies there is normally lower birth weight and this and other factors that affect growth of the foetus *in utero* can be significant risk factors for development of reproductive cancers in both sexes. This is illustrated in Study 13, in which risk of breast cancer in a female twin was considerably increased when there was a male twin present, and this appeared to be related partly to an increase in birth weight of the female twin. The latter effect might be related to increased androgen exposure from the male foetus, as the female foetus makes negligible amounts of sex steroids.

1.3.6.2 WG Conclusions

While the WG felt it reasonable to conclude that the hormone environment *in utero* is a factor in determining subsequent risks of some reproductive cancers, this is a complex area to interpret. It is certainly not straightforward to conclude on the basis of these findings that pre-natal exposure of the foetus to exogenous hormones, in particular hormones used as growth promoters in livestock, will be capable of inducing comparable effects. Issues such as potency, bioavailability, pharmacokinetics and transfer to the foetus all have to be taken into account.

By reference to offspring from women who were treated with extremely high doses ($>0.1\,mg\,kg^{-1}day^{-1}$) of the potent oestrogen diethylstilboestrol during pregnancy, only a very modest increase in testicular cancer risk occurred in the male offspring[66] and only rare cases of vaginal cancer occurred in the female offspring.[67] It would therefore seem unlikely that exposure to the less potent growth-promoting compounds, at what would be very much lower concentrations,[12–14] would pose a significant risk with regard to the development of reproductive cancers. Moreover, experimental studies in rodents that involve administration of test compounds to pregnant animals and consequent exposure of the foetus may be poor models for the human, because of major differences in endogenous hormone concentrations, timing/duration of foetal development,[63] *etc.* Again, the dramatic changes in diet, BMI, later age at first pregnancy, rates of smoking, *etc.* in women in Western countries over the past 50 years have established effects on foetal growth and development.[11] Against this changing background, discerning potential contributory effects from low-level exposure to growth-promoting hormones or their metabolites is probably an impossible task.

1.3.7 Environmental Impact of Hormonally Active Substances

1.3.7.1 Recent Data

Although the SCVPH Opinion (2002) concentrated on risks to human health, Section 6 (p. 21) and Annex 1 (pp. 24–27) also considered environmental effects. Three studies are mentioned by the SCVPH in relation to environmental effects:

Endocrine disrupting activity of anabolic steroids used in cattle. The paper by Schiffer *et al.* (2001)[68] contains results from this study that are of relevance to environmental risk assessment.

Screening water samples for estrogenic and androgenic anabolic chemicals. The results from this study had not been fully published, but a discussion is published in a brief paper by Jégou *et al.* (2001).[69]

Endocrine disrupting effects of cattle farm effluent on environmental sentinel species. The results from this study have been published in brief form in a review by Orlando and Guillette (2001).[70]

The WG evaluated the SCVPH's conclusions and compared these with the published evidence from the three cited studies. Additional information was not sought, and it is possible that further publications have emerged from the three SCVPH studies.

Section 6 of the SCVPH Opinion simply stated that previous SCVPH Opinions have not addressed environmental concerns, but that relevant results from the 17 Studies are presented in Annex 1. This section also draws attention to the existence of a report by the Scientific Committee on Toxicity, Ecotoxicity and the Environment.[71] This report is an overview of the evidence for endocrine disruption, particularly in wildlife. It does not specifically address risks from hormone residues in beef, but does briefly consider human and wildlife studies that examine the effects of oestradiol and 17α-ethinyloestradiol. The main focus for wildlife studies is on chlorinated organic compounds (*e.g.* PCBs and DDT) and TBT, rather than hormones. Several recommendations are made to improve the predictive and monitoring tools for detecting endocrine disrupting chemicals that might occur in the environment. However, the direct relevance of this CSTEE report to the environmental risk assessment of hormones used in beef is rather limited.

The SCVPH Opinion Annex 1 reviews the three studies with relevance for the environment, and drew five principal conclusions:

- Aquatic animals are most sensitive to endocrine disruptors due to a greater potential for tissue accumulation.
- The environmental impact of anabolic steroids is potentially great.
- Further studies to determine the biological and chemical stability of such steroids in soil and water are warranted.
- Little information is available on the endocrine disrupting potential of the metabolites of MGA.
- Surface water downstream from a cattle feedlot was contaminated with oestrogenic and androgenic compounds, but the identity of these could not be established. Fish morphology near to cattle feedlots showed signs of endocrine disruption but, once again, the specific cause of this could not be identified.

The WG considered these five conclusions in turn, as follows:

(i) *Aquatic animals are most sensitive to endocrine disruptors due to a greater potential for tissue accumulation.* Two issues were thought to be confused here: the inherent sensitivity of an organism to a toxicant and the extent to which organisms in different environmental compartments (*e.g.* freshwater, seawater, land or air) may be exposed to these contaminants. This conclusion requires further work in two areas before it can be accepted. There needs to be further ecotoxicological testing of the relative sensitivity of different terrestrial and aquatic organisms to growth-promoting hormones used in beef production. There also needs to be further environmental chemistry to determine the pathways taken by these hormones once they are released into the environment, to examine whether it is likely that they will reach aquatic systems.

(ii) *The environmental impact of anabolic steroids is potentially great.* This conclusion was based upon the findings from Schiffer *et al.* (2001)[68] on the degradation kinetics of excreted trenbolone acetate (TBA) and melengestrol acetate (MGA) under different manure storage conditions. The study showed that both hormones are excreted in faeces and can be detected in soil for up to several months when contaminated dung that has been stored for 4.5 to 5.5 months is applied. There was evidence that both trenbolone and MGA adsorb strongly to soil. The authors speculated that various physical or biological processes could eventually remove these hormones from soil, but no work was done to determine which, if any, of these removal processes is most likely. This was a well-performed study, but it did not demonstrate, or seek to demonstrate, that either of these hormones had an adverse impact on the environment. It simply demonstrates that there is a pathway for these hormones from beef cattle, through dung and into soil. A pathway from soil into either terrestrial or aquatic organisms and subsequent biological effects in these organisms would need to be demonstrated before one could state that there is an environmental impact.

(iii) *Further studies to determine the biological and chemical stability of such steroids in soil and water are warranted.* This conclusion is also based on Schiffer *et al.* (2001)[68] and agrees with Schiffer *et al.*'s conclusions. This is an appropriate conclusion to reach; clearly the conclusion stated in (ii) above cannot be supported until these stability studies are done.

(iv) *Little information is available on the endocrine disrupting potential of the metabolites of MGA.* This is also based on Schiffer *et al.* (2001).[68]

(v) *Surface water downstream from a cattle feedlot was contaminated with oestrogenic and androgenic compounds, but the identity of these could not be established. Fish morphology near to cattle feedlots showed signs of endocrine disruption but, once again, the specific cause of this could not be identified.* These views were thought to be based on a study which has apparently not been published in the peer-reviewed literature, except for a brief summary in Jégou *et al.* (2001).[69] The conclusions are also based on a study, which has only been published briefly as part of a review.[70] These published papers do not provide sufficient information to judge the quality of the work, although the researchers involved are acknowledged leaders in the field.

1.3.7.2 WG Conclusions and Recommendations

The three environmental studies cited by the SCVPH Opinion were important initial efforts to understand the environmental risks posed by use of growth-promoting hormones in beef production. However, it is clear, as acknowledged by SCVPH, that these studies do not provide strong evidence that growth-promoting hormones used in beef production are the cause of oestrogenic and androgenic activity in water below feedlots, or of de-masculinisation of fish. Application of Hill's criteria or Koch's postulates to these results suggests that much more work needs to be done before uncertainties are reduced to a degree at which an evidence-based decision can be made. In particular, the identity of the substances responsible for such effects needs to be established, probably through Toxicity Identification Evaluation procedures, and a more extensive set of sites should be investigated so that problems of pseudo-replication are avoided.

One study had been reported fully in the peer-reviewed literature and showed that at least two of the hormones, or their metabolites, may be found in soil after dung spreading. However, pathways from soil (where these compounds may be strongly bound and therefore unavailable) to sensitive biological receptors, have not been established. In contrast to this study, the rather limited published results from the other two investigations showed that endocrine disrupting substances seem to be present, and may be exerting biological effects, at river sites near to cattle feedlots. However, the substances responsible for these effects have not been identified, so a pathway from effects on environmental receptors to application of hormones in beef remains un-established.

The WG believed results from these studies are insufficient to demonstrate cause and effect. Research to establish a source–pathway–receptor linkage is required. Importantly, the WG assumed that if the re-introduction of growth-promoting substances for use within the EU were considered in the future, then a full environmental risk assessment would need to be conducted according to good current scientific practice.

1.3.8 Other Considerations of the WG

1.3.8.1 Formal Risk Assessment of Hormonally Active Substances

The WG discussed whether it would be possible to undertake a formal risk assessment on hormonally active compounds from the available data on residue concentrations and consumption data. The individual substances of concern all have ADIs and thus, in theory, it would have been possible to compare any estimated dose with the current ADIs. To this end, the UK Food Standards Agency is able to provide reliable data on food consumption for toddlers and adults that include dairy, meat and aquaculture products. The information on residue concentrations for the growth-promoting substances (whether natural or synthetic) in bovine meat and meat products, together with the data on UK food consumption, would in the future enable a total body dose to be calculated should this be required. This would require a better understanding of residue concentrations from both proper and improper use of the hormonal substances and, in the light of the newer scientific information, whether or not new ADIs might be required.

It is arguably also important to distinguish the natural hormones, which, if present in food, simply supplement those already circulating in the body, from the synthetic analogues which may have subtle differences in receptor binding, metabolism, *etc*. For natural hormones it could be argued that the traditional approach of deriving ADIs by applying safety factors to no observed adverse effect levels (NOAELs) is excessively conservative. The risks following ingestion of natural hormones in food which simply supplement those already circulating in the body can only truly be assessed from human data and should be put into context with the intake from other sources, the dose relative to normal circulating concentrations and the human evidence of the adverse effects of elevated circulating hormone concentrations. If the evidence suggests that any chronic increase in oestrogen concentrations will tend to increase the risk of breast cancer, as seems to be the case, we need to derive a view of an acceptable degree of increase in that risk, and prioritise the sources of exposure for control against that value. Essentially this might mean advice to avoid any contribution of oestrogens from the diet for anyone already taking supplementary oestrogens.

The WG therefore believed that there is a need to gain a much better understanding of the impact of very small increases in oestrogen concentrations in human populations. If it is demonstrated that the contribution from meat from oestrogen-treated animals is not increased above that from untreated animals then the actual treatment is an irrelevance in the debate.

Synthetic hormones (and lipoidal esters) cannot be treated in the same way, since we do not have enough information about their interaction with natural hormones in humans, and probably the only way to arrive at an estimate of a safe human dose is to obtain more data in experimental animals. It would also be important to attempt to model the interaction of synthetic hormones with natural oestrogens in order to explore the possibility of their increasing or decreasing the incidence of tumours in humans.

1.3.8.2 Ban on Over-thirty-month Cattle

The WG considered the effect of the end of the ban in the UK in 2005 on cattle over 30 months of age entering the food chain. It noted that this would lead to a massive increase in the endogenous concentrations of oestradiol in meat reaching the consumer; although it was also noted that this increase would be to pre-ban concentrations of nine years earlier. This in itself would likely dwarf any increase in hormone concentrations reaching the consumer as a result of the use of growth-promoters, were they to be re-introduced. However, set against this consideration was the argument that, if 17β-oestradiol increases the risk of breast cancer as previously discussed, then any avoidable increase, however small, would have to be viewed as undesirable.

The view was also expressed by the WG that other factors beyond our control are likely to (and probably already do) have a bigger effect on hormone concentrations in food than the use of growth promoters, *e.g*. a change in the type of sire used in the national dairy herd. The point was also raised that, if we were to continue to exclude growth promoters on the grounds that it would raise the exposure of hormones to the consumer, then logically one should consider excluding meat from pregnant animals and possibly those in oestrus.

JECFA[15] concluded in their evaluation of the numerous studies using authorised doses of the three natural steroids either alone or in combination that the hormone concentrations in edible tissues and blood were sometimes statistically significantly higher than the corresponding values found in concurrent controls but were always within the physiological range of these substances in bovine animals. The highest concentrations of progesterone and 17β-oestradiol are found in lactating and pregnant cows whereas extremely high concentrations of testosterone are found in bulls; the concentrations in treated calves and steers are significantly less than these natural concentrations.

The excess contribution of the residues to the ADIs set by JECFA is <4% for oestrogens, approximately 0.003% for progesterone and 0.2% for testosterone. Bearing in mind that post-pubertal humans produce very much larger quantities of these hormones, the margin of safety for adults consuming meat from treated animals is very high. The pre-pubertal child produces fewer natural steroids but will always consume fewer than the respective ADIs.

1.3.8.3 Conclusions and Recommendations of the WG

As can be seen from the above information and discussions, the WG gave very careful consideration to both the views of the SCVPH Opinion as well as the published papers and reports coming out of the 17 EC-funded studies. In addition, the WG included in their deliberations other papers and discussions which helped to put the use of these hormonally active substances into a broader context so as to assist risk assessors, policy-makers and regulators. The conclusions of the WG are reproduced below in their entirety.

1.3.8.4 Current Scientific Evidence For or Against Adverse Effects

The previous sections have discussed the current new evidence relating to mutagenicity, carcinogenicity and endocrine disrupting effects of the hormonally active substances and for humans who may be consuming meat from treated animals. Most of the evidence relates to 17β-oestradiol and the following key features are considered relevant to this and any future hazard or risk assessment:

- 17β-estradiol can be activated to catechol oestrogens and then oxidised to form semiquinones. The metabolites may also generate potentially mutagenic oxygen radicals by redox cycling.
- There is good evidence for the formation of DNA adducts from metabolites of 17β-oestradiol *in vitro* and *in vivo*.

- Synthesised oestrogen metabolite adducts are pre-mutagenic in sub-cellular test systems.
- There is some evidence for mutagenic potential (induction of chromosome damage and mutations) for some metabolites of oestradiol in mammalian cells *in vitro*.
- The evidence for induction of chromosome aberrations and gene mutations *in vivo* is poor and is derived from non-standard studies.
- There are, however, reasonable arguments against the carcinogenicity of 17β-oestradiol being due solely to epigenetic processes.
- It would be prudent to consider oestradiol and its metabolites as a complete carcinogen whilst more substantial evidence for its mode of action is obtained.
- Despite its possible genotoxicity, it is reasonable to consider that 17β-oestradiol may have a threshold for carcinogenicity due to the presence of homeostatic feedback mechanisms, the requirement for activation pathways to exceed inactivation pathways and the presence of antioxidants *in vivo*.

When it came to the current evidence base for 17β-oestradiol, however, in spite of certain data gaps, the view of most of the Working Group was that there is ample information to show that zootechnical and therapeutic uses of 17β-oestradiol do not pose any risk to humans unless an active implant site is ingested.

In relation to the other hormones considered, a number of points emerged in the Working Group discussions. One view was that in regard to the five other hormones (testosterone, progesterone, trenbolone, zeranol and MGA), one could agree with the SCVPH assessment, as expressed in Directive 2003/74/EC of the European Parliament and the Council of 22 September 2003, "that the current state of knowledge does not make it possible to give a quantitative estimate of the risk to consumers". However, the majority of the Working Group felt that, in spite of the acknowledged data gaps and uncertainties, the available evidence on genotoxicity, tumorigenicity, hormonal activity and endocrine disrupting effects was supportive of the view that eating meat from animals treated with these five hormones was unlikely to be harmful to human health.

As a rider to these statements, it should be noted that they are based on the assumption that the consumer is exposed to no greater concentrations of residues than those arising from "correct" or "recommended" use of hormones. The likely misuse of growth-promoting substances is noted elsewhere in this report.

A number of additional points were made by members of the Working Group:

- In spite of the 17 additional studies funded by the Commission, little progress has been made to determine the safety of hormone growth promoters.

- One member felt strongly in support of the findings of the SCVPH – that much more work needs to be done before the safety of the six substances under consideration can be assured and approval given for their growth-promotional purposes.
- At the time, the 1999 VPC Sub-Group reported, they were unable to support the conclusions reached by the SCVPH "that risks associated with the consumption of meat may be greater than previously thought". However, whilst the last 1999 VPC Sub-Group reported that "none of the publications reviewed provide any substantive evidence that oestradiol was mutagenic/genotoxic", the more recent evidence does indicate that:
  - a) metabolites of oestradiol do have the potential to be genotoxic, *in vitro* and *in vivo*; and
  - b) steroid metabolites previously considered to be nothing more that inactivation products may have patho-physiological actions themselves.

1.3.8.5 Overall Conclusions and Recommendations

1. The Working Group were of the view that human exposure to residues of hormonally active substances, including growth promoters in meat, could exert biological effects if exposure is at a sufficiently high level. Therefore, the two key issues are:
   - (i) determination of the dose-response induction of biological effects by the hormonally active substances in test animals and, ideally, humans in order to identify a Lowest Observable Effect Level (LOEL), and
   - (ii) determination of the level (and range) of the additional human exposure and uptake from eating meat from treated animals.
2. These determinations should be made in adults and in developing (foetal/neonatal) animals and humans to identify the most sensitive index of effect. These effects would be in addition to those occurring naturally due to endogenous hormones.
3. The research so far has provided some, but not all, the basic, but essential, information outlined above. Without it, no definitive conclusions can be drawn; although the weight of available evidence suggests that likely levels of human exposure to hormonally active substances in meat from treated animals would not be sufficient to induce any measurable biological effect.
4. Specifically, it is very unlikely that the presence of 17β-oestradiol and its metabolites in meat from treated animals would significantly increase the risk of adverse effects in consumers. This is due to their low concentrations in comparison to those arising from endogenous production and from other dietary sources. Any increase would be likely to be small in the context of the whole food basket.

5. In reaching these conclusions, the Working Group expressed a number of qualifications and reservations based on the current lack of evidence of a risk to humans. These included:
   - all scientific judgements made by the Working Group were based on the assumption that the consumer is exposed to residues at no greater concentrations than those that would be caused by the "correct" or "recommended" use of the exogenous hormones, be it for growth promotion or other zootechnical uses or therapeutic purposes;
   - the Working Group understands that misuse of hormonally active substances for growth promotion is more likely than misuse for oestrus synchronisation or therapeutic uses; and
   - substances with hormonal action may be used in combination, both legally and illegally, while the toxicological and safety factors available (*e.g.* MRLs and ADIs) only relate to single substances.
   - the Working Group had to decide what to do in the absence of information or where there was uncertainty of interpretation of information. One member expressed the view that, for the substances under consideration, there was a large element of uncertainty, so the precautionary principle must assume the primary consideration. The many uncertainties associated with the current lack of knowledge could be addressed by further research where this was both feasible and affordable. The Working Group was unanimous that all uncertainties must be made clear, especially those that were considered crucial in the risk assessment process.
6. As has been noted in this report, and acknowledged in the SCVPH Opinion, there are important gaps in the evidence base that preclude producing definitive risk assessments for 17β-oestradiol or the other five hormonally active substances. Not all data gaps are equally important for the purposes of risk assessment and the Working Group highlighted a number that could improve future risk assessments. As an example, it would be helpful if the CVMP and JECFA could make available data on pharmacokinetics and metabolism of assessed compounds that were supplied in manufacturers' dossiers. This openness and transparency would allow greater public scrutiny of the facts and confidence in the hazard and risk assessments produced.
7. The Working Group felt that none of the basic issues could be addressed without a structured approach. There is a need to establish precisely the:
   - relationships between the potential use of growth promoters (including over-use) and concentrations of residues in meat;
   - levels of exposure in consumers (*i.e.* taking account of intake, absorption, bioavailability and metabolism); and
   - dose-response relationships for the effects of the hormonally active substances (and their metabolites) in experimental animals or in humans.

  - Further data on lipoidal oestrogens, possible bioaccumulation and possible synergistic effects of cocktails of hormonal substances would also be desirable.
8. The Working Group noted specific needs:
  - To establish in humans a detailed dose-response curve that relates exposure to specific hormonally active substances to the amount of meat consumed from treated animals.
  - To establish in experimental animals the relationship between intake of hormonally active substances, or their metabolites, and target-organ effects (selecting the likely most sensitive target organ depending on the nature of the activity of the compound). This study to be conducted for adults and then foetal and/or neonatal exposure to be considered.
  - To consider lipoidal esters of oestrogen in future studies of the possible passage of oestrogen in implants through cattle to humans. The bioavailability and metabolism of lipoidal esters following ingestion should be investigated to allow the biological significance of the oestrogens to be assessed.
  - To carry out studies to confirm whether the ADI for pre-pubertal boys could be exceeded if they consumed a standard 500 g portion of meat from an animal that had been treated with a number of hormonal implants. If confirmed this would be of concern.
  - To establish an independent laboratory test to confirm that meat has not been derived from animals produced with the aid of growth-promoting hormone implants.

1.4 Recent Opinions and the Future

Since the publication of the WG 2005 report, there have been a number of publications relevant to all of the scientific areas discussed above. Of particular importance is a new Opinion of the Scientific Panel of Contaminants in the Food Chain,[72] which reviewed all relevant papers published between 2002 and 2007 in relation to testosterone, progesterone, trenbolone acetate, zeranol and melengestrol acetate used as growth promoters in meat production. The request for the Opinion came from the European Commission to the European Food Standards Authority (EFSA) to which the SPCFC belong. The terms of reference were somewhat similar to that of the VPC WG 2005 but a little more limited in that they were only required to summarise to new data. The conclusions that they reached are reproduced below.

1.5 Conclusions

- New data published since 2002 confirm and extend the current understanding of the effects of steroid hormones and hormone-like substances used as growth-promoting hormones (GPHs), which are not only *via* interactions with their specific receptors.

- In *in vitro* systems, the potencies of zeranol, trenbolone and melengestrol acetate in terms of oestrogen, androgen and progesterone receptor affinities and modulation of gene expression, as well as cell proliferation and apoptosis, may be equal to, or exceed, those of the most active natural hormones. There is a lack of information with respect to the *in vivo* significance of these effects at exposure levels associated with residues in meat.
- Sensitive analytical methods have become available permitting the identification and quantification of the growth-promoting hormones (all five compounds under consideration) and their currently known major metabolites. Regarding the natural hormones, testosterone and progesterone, these methods also allow discrimination between endogenous and exogenous hormone residues. These advanced methods have as yet only been used in a very limited number of experimental studies, and await to be applied on a broader scale.
- In the absence of data from surveillance studies, the exposure to residues of the hormones used as growth-promoting agents cannot be quantified. In particular, the available data on the metabolism of trenbolone, zeranol or melengestrol acetate in cattle, and the amount and nature of residues in animal tissues following routine use of these compounds in beef cattle operations, are too incomplete to be assessable.
- An increasing number of publications presenting epidemiological data indicate a correlation between red-meat consumption and hormone-dependent cancers of the breast and prostate. Due to the high number of confounding factors, the contribution of residues of hormones in meat cannot be quantified in these studies.
- Large-scale cattle production and the use of growth-promoting hormones in cattle operations in Third Countries has been associated with undesirable effects in sentinel aquatic species in contact with cattle farm effluents.

The CONTAM Panel concluded that the new data that are publicly available do not provide quantitative information that would be informative for risk characterisation, and therefore do not call for a revision of the previous assessments of the Scientific Committee on Veterinary Measures relating to Public Health (SCVPH).

As can be seen, there is nothing fundamentally new in these conclusions that contradicts those of the VPC WG 2005 in relation to the specific substances. Of note however, the final paragraph seems to support the risk assessment of the previous SCVPH 1999, 2002 Opinions that were used to underpin and maintain the EU ban.

Of scientific importance is that the better analytical methods have been developed which are able to distinguish between endogenous and exogenous sources for the compounds of concern in tissue residues. These are said to be of only limited utilisation and would need further development for routine residue surveillance schemes.

Critically, all experts groups seem to agree that the available scientific and epidemiological data is insufficient to come to firm conclusions regarding risk assessments for the compounds of concern, for humans consuming meat, for any of the health endpoints.

Clearly, this story is not over. The research will continue and new Opinions will be provided by expert committees at the requests of regulatory bodies. This is not surprising as the health consequences are serious (cancer and reproductive perturbations) and the major proportion of the human population will be exposed to meat and meat products where use of these substances as growth promoters is permitted.

References

1. SCVPH, Scientific Committee on Veterinary Measures Relating to Public Health, Opinion of the Scientific Committee on Veterinary Measures Relating to Public Health: Assessment of potential risks to human health from hormone residues in bovine meat and meat products (30 April 1999), http://europa.eu.int/comm/food/fs/sc/scv/out21_en.html, 1999, accessed 29 March 2009.
2. JECFA, Joint FAO/WHO Expert Committee on Food Additives, Fifty-second meeting, Rome, 2–11 February, 1999 [published as WHO, 2000].
3. SCVPH, Scientific Committee on Veterinary Measures Relating to Public Health, Review of Specific Documents Relating to the SCVPH Opinion of 30 April 99 on the Potential Risks to Human Health from Hormone Residues in Bovine Meat and Meat Products, http://europa.eu.int/comm/food/fs/sc/scv/out33_en.pdf, 2000, accessed 29 March 2009.
4. SCVPH, Scientific Committee on Veterinary Measures Relating to Public Health, Review of Previous SCVPH Opinions of 30 April 1999 and 3 May 2000 on the Potential Risks to Human Health from Hormone Residues in Bovine Meat and Meat Products (adopted on 10 April 2002), http://europa.eu.int/comm/food/fs/sc/scv/out50_en.pdf, 2002, accessed 29 March 2009.
5. WHO, Global Assessment of the State-of-the-Science of Endocrine Disruptors (International Programme on Chemical Safety, WHO/PCS/EDC/02.2), Geneva, Switzerland, World Health Organization, http://www.who.int/ipcs/publications/new_issues/endocrine_disruptors/en, 2002, accessed 29 March 2009.
6. T. M. Crisp, E. D. Clegg, R. L. Cooper, W. P. Wood, D. G. Anderson, K. P. Baetcke, J. L. Hoffmann, M. S. Morrow, D. J. Rodier, J. E. Schaeffer, L. W. Touart, M. G. Zeeman and Y. M. Patel, *Environ. Health Perspect.*, 1998, **106**(1), 11–56.
7. H. M. Bolt, P. Janning, H. Michna and G. H. Degen, *Arch. Toxicol.*, 2001, **74**, 649–662.
8. COT (Committee on Toxicity of Chemicals in Food, Consumer Products and the Environment), Phytoestrogens and health. Foods Standards

Agency, London, http://cot.food.gov.uk/cotreports/cotwgreports/phytoestrogensandhealthcot, 2003, accessed 29 March 2009.
9. G. D. Hammond, *Obstet. Gynecol. Clin. North Am.*, 2002, **29**, 411–423.
10. V. C. Jordan, *Sci. Amer.*, 1998, **279**, 60–67.
11. R. M. Sharpe and S. Franks, *Nat. Med.*, 2002, Supplement 10, s33–s40.
12. D. M. Henricks, S. L. Gray, J. J. Owenby and B. R. Lackey, *APMIS*, **109**, 273–283.
13. I. G. Lange, A. Daxenberger and H. H. D. Meyer, *APMIS*, 2001, **109**, 53–65.
14. D. Maume, Y. Deceunick, K. Pouponneau, A. Paris, B. Le Bizec and F. Andre, *APMIS*, 2001, **109**, 32–38.
15. JECFA, Residues of some veterinary drugs in animals and food, Fifty-second meeting, Estradiol-17β, Progesterone and Testosterone, FAO Food and Nutrition Paper 41/12, 1999, pp. 37–90.
16. G. H. Singleton and H. Dobson, *Vet. Record*, 1995, **136**(7), 162–165.
17. J. M. Scheffler, D. D. Buskirk, S. R. Rust, J. D. Cowley and M. E. Doumit, *J. Anim. Sci.*, 2003, **81**, 2395–2400.
18. T. L. Mader, Carry-over and lifetime effects of growth promoting implants, OSU Conference: Proc. Impact of Implants on Performance and Carcass Value of Beef Cattle, 1997, pp. 88–94.
19. W. J. Platter, J. D. Tatum, K. E. Belk, J. A. Scanga and G. C. Smith, *J. Anim. Sci.*, 2003, **81**, 984–996.
20. S. K. Duckett and J. G. Andrae, *J. Anim. Sci.*, 2000, **79**(E), 110–117.
21. B. L. Barham, J. C. Brooks, J. R. Blanton, A. D. Herring, M. A. Carr, C. R. Kerth and M. F. Miller, *J. Anim. Sci.*, 2003, **81**, 3052–3056.
22. B. Le Bizec, P. Marchand and F. Andre, *Annales de Toxicologie Analytique*, 2000, **XII**(1), 56–63.
23. P. Marchand, B. le Bizec and C. Gade, *et al., J. Chrom.*, 2000, **867**, 219–233.
24. R. Le Guevel and F. Pakdel, *Hum. Reprod.*, 2001, **16**, 1030–1036.
25. L. A. P. Hoogenboom, Investigation on the metabolism of 17â-estrodiol by bovine hepatocytes, human intestinal and breast cells and the genotoxic and estrogenic properties of the metabolites, unpublished report, RIKILT Research Institute, 2000.
26. L. A. P. Hoogenboom, L. De Haan, D. Hooijerink, G. Bor, A. J. Murk and A. Brouwer, *APMIS*, 2001, **109**, 101–107.
27. S. E. Hankinson, J. E. Manson, D. Spiegelman, W. C. Willett, C. Longcope and F. E. Speizer, *Canc. Epidemiol. Biomarkers Prev.*, 1995, **4**(6), 649–654.
28. V. Beral, *Lancet*, 2003, **362**, 419–427.
29. R. T. Chlebowski, S. L. Hendrix, R. D. Langer, M. L. Stefanick, M. Gass, D. Lane, R. J. Rodabough, M. A. Gilligan, M. G. Cyr, C. A. Thomson, J. Khandekar, H. Petrovitch and A. McTiernan, *J. Am. Med. Assoc.*, 2003, **289**, 3243–3253.
30. S. Yen, P. Martin, A. Burnier, N. Czekala, M. Greaney and M. Callantine, *J. Clin. Endocrinol. Metabol.*, 1975, **40**, 518–521.

31. D. Arnold, in Residues of some veterinary drugs in animals and foods, 52nd JECFA meeting, FAO Food and Nutrition Paper 41-12, ftp://ftp.fao.org/es/esn/jecfa/vetdrug/41-12-estradiol_17%DF_progresterone_testoterone.pdf, 1999, accessed 29 March 2009.
32. I. E. Smith and M. Dowsett, *New Engl. J. Med.*, 2003, **348**(24), 2431–2442.
33. M. M. Braun, A. Ahlbom, B. Floderus, L. A. Brinton and R. N. Hoover, *Canc. Causes Contr.*, 1995, **6**, 519–524.
34. A. J. Swerdlow, B. L. De Stavola, M. A. Swanwick and N. E. S. Maconochie, *Lancet*, 1997, **350**, 1723–1728.
35. B. Kappel, K. Hansen, J. Moller and J. Faaborg-Andersen, *Acta Genet. Med. Gemello (Roma)*, 1985, **34**, 99–106.
36. M. Kaijser, P. Lichtenstein, F. Granath, G. Erlandsson, S. Cnattingius and A. Ekbom, *J. Natl. Cancer Inst.*, 2001, **93**(1), 60–62.
37. M. Feldman, C. T. Richardson and J. H. Welsh, *J. Clin. Invest.*, 1983, **71**, 715–720.
38. C. J. Hawkey, I. Wilson, J. Naesdal, G. Langstrom, A. J. Swannell and N. D. Yeomans, *Gut*, 2002, **51**, 344–350.
39. L. Chang and M. M. Heitkemper, *Gastroenterology*, 2002, **123**, 1686–1701.
40. J. J. DeCosse, S. S. Hgoi, J. S. Jacobson and W. J. Cennerazzo, *Eur. J. Cancer*, 1993, **Prev. 2**, 105–115.
41. M. Campbell-Thompson, K. K. Reyher and L. B. Wilkinson, *J. Endocrinol*, 2001, **171**, 65–73.
42. J. Cosnes, F. Carbonnel, F. Carrat, L. Beaugerie and J. P. Gendre, *Gut*, 1999, **45**, 218–222.
43. H. Leffers, M. Naesby, B. Vendelbo, N. E. Skakkebaek, M. Jorgensen, P. Grandjean, W. Sippell, A. Soto, G. Vollmer and H. Meyer, *Hum. Reprod.*, 2001, **16**(5), 1037–1045.
44. C. Mastri, P. Mistry and G. W. Lucier, *J. Steroid Biochem.*, 1985, **23**, 279–289.
45. K. Shrimanker, L. J. Salter and R. L. Patterson, *Horm. Metab. Res.*, 1985, **17**, 454–457.
46. Z. Ben-Rafael, L. Mastroianni Jr, F. Meloni, M. S. Lee and G. L. Flickinger, *J. Clin. Endocrinol. Metab.*, 1986, **63**, 1106–1111.
47. S. C. Nagel, F. S. vom Saal and W. V. Welshons, *Proc. Soc. Exper. Biol. Med.*, 1998, **217**, 300–309.
48. D. G. Lindsay, *Fd. Chem. Toxic.*, 1985, **23**(8), 767–774.
49. E. Cavalieri, K. Frenkel, J. G. Liehr, E. Rogan and D. Roy, *J. Natl. Cancer. Monographs*, 2000, **27**, 75–93.
50. J. A. Lavigne, J. E. Goodman, T. Fonong, S. Odwin, P. He, D. W. Roberts and J. D. Yager, *Cancer Res.*, 2001, **61**, 7488–7494.
51. J. C. Liehr, *Hum. Reprod. Update*, 2001, **7**, 273–281.
52. I. Terashima, N. Suzuki and S. Shibutani, *Biochemistry*, 2001, **40**, 166–172.
53. T. Tsutsui, Y. Tamura, M. Hagiwara, T. Miyachi, H. Hikiba, C. Kubo and J. C. Barrett, *Carcinogenesis*, 2000, **21**, 735–740.
54. T. Tsutsui, Y. Tamura, E. Yagi and J. C. Barrett, *Int. J. Cancer*, 2000, **86**, 8–14.

55. L.-Y. Kong, P. Szaniszlo, T. Albrecht and J. G. Liehr, *Int. J. Oncology*, 2000, **17**, 1141–1149.
56. D. Chakravarti, O. C. Mailander, K.-M. Li, S. Higginbotham, H. L. Zhang, M. L. Gross, J. L. Meza, E. L. Cavalieri and E. G. Rogan, *Oncogene*, 2001, **20**, 7945–7953.
57. E. Yared, T. J. McMillan and F. L. Martin, *Mutagenesis*, 2002, **17**, 345–352.
58. M. Richold, *Arch. Toxicol.*, 1988, **61**, 249–258.
59. M. Metzler and E. Pfeiffer, *APMIS*, 2001, **109**, 89–95.
60. E. Yagi, J. C. Barrett and T. Tsutsui, *Carcinogenesis*, 2001, **22**, 1505–1510.
61. I. G. Lange, A. Daxenberger, H. H. D. Meyer, E. Rajpert-De Meyts, N. E. Skakkebaek and D. N. R. Veeramachaneni, *Xenobiotica*, 2002, **32**(8), 641–651.
62. G. Chiumello, M. P. Guarneri, G. Russo, L. Stroppa, P. Sgaramella, M. Joffe, P. Thonneau, A.-M. Andersson, P. Myers, J. Toppari, J. Huff, S. De Muinck Keizer-Schrama, H. Kulin, A. Daxenberger and J. P. Bourguignon, *APMIS Supplementum*, 2001, **109**(103), S203–S209.
63. R. J. Witorsch, *Fd. Chem. Toxicol.*, 2002, **40**, 905–912.
64. T. Tuvemo and L. A. Proos, *Ann. Med.*, 1993, **25**, 217–219.
65. R. Virdis, M. E. Street, M. Zampolli, G. Radetti and B. Pezzini, *et al., Arch. Dis. Child.*, 1998, **78**, 152–154.
66. J. Toppari, J. C. Larsen, P. Christiansen, A. Giwercman and P. Grandjean, *et al., Environ. Health Perspect.*, 1996, **104**(4), 741–803.
67. K. L. Noller, *Cervix Low. Female Gen. Tract*, 1983, **1**, 75–82.
68. B. Schiffer, A. Daxenberger, K. Meyer and H. H. D. Meyer, *Environ. Health Perspect.*, 2001, **109**, 1145–1150.
69. B. Jégou, A. Soto, S. Sundlof, R. Stephany, H. Meyer and H. Leffers, *APMIS*, 2001, **109**(103), S551–S556.
70. E. F. Orlando and L. J. Guillette Jr, *Hum. Reprod. Update*, 2001, **7**, 765–272.
71. CSTEE, CSTEE Opinion on Human and Wildlife Health Effects of Endocrine Disrupting Chemicals, with Emphasis on Wildlife and on Ecotoxicology Test Methods. Report of the Working Group on Endocrine Disrupters of the Scientific Committee on Toxicity, Ecotoxicity and the Environment (CSTEE) of DG XXIX, Consumer Policy and Consumer Health Protection, European Commission, March 1999.
72. SPCFC, http://www.efsa.europa.eu/cs/BlobServer/Scientific_Opinion/CONTAM_ej510_hormone_op_en.pdf?ssbinary=true, 2007, accessed 29 March 2009.

CHAPTER 2

Presence and Metabolism of Endogenous Steroid Hormones in Meat-producing Animals

JAMES SCARTH[a] AND CHRISTINE AKRE[b]

[a] HFL Sport Science (A Quotient Bioresearch company), Newmarket Road, Fordham, CB7 5WW, Cambridgeshire, United Kingdom; [b] Canadian Food Inspection Agency, 116 Veterinary Road, Saskatoon, Canada, S7N 2R3

2.1 Introduction

The Oxford English Dictionary defines hormones as "any of numerous organic compounds that are secreted into the body fluids of an animal, particularly the bloodstream, by a specific group of cells and regulate some specific physiological activity of other cells; also, any synthetic compound having such an effect".[1] Merriam-Webster further defines sex hormones as "a steroid hormone (as oestrogen or testosterone) that is produced especially by the ovaries, testes, or adrenal cortex and affects the growth or function of the reproductive organs or the development of secondary sex characteristics".[2] They are considered to be organ specific rather than gender specific and influence many tissues and organs in the body as well as the sex organs.[3] Nearly all cells can be regarded as target tissues for sex hormones and as such they play a very important role in the maintenance and balance of the body at the cell level and can affect many of the physiological processes that occur. The metabolism of these compounds is

RSC Food Analysis Monographs No.8
Analyses for Hormonal Substances in Food-producing Animals
Edited by Jack F. Kay

Published by the Royal Society of Chemistry, www.rsc.org

essential for moving them around the body, modifying their activity and aiding in excretion.

The European Union (EU) Council Directive 96/22/EC of 1996[4] states that " . . . substances having a hormonal action . . . " are prohibited for use in animals intended for meat production". These compounds are defined under Group A of Annex I of the Council Directive 96/23/EC[5] as substances having an anabolic effect; unauthorised substances, including natural or synthetic steroids; stilbenes, their derivatives, salts and esters; antithyroid agents; resorcyclic acid lactones, including zeranol; β-agonists and compounds included in Annex V to Council Regulation no. 2377/90/EC.[6] The same Council Directive (and Council Decision 2002/657/EC[7]) then lay down requirements for residue testing in order to ensure compliance with the prohibition. A succinct overview of the context of these different directives and the resulting analytical methods that have been applied for these and other substances in recent years can be found in Stolker *et al.*[8] Indeed, existing EU guidelines for positive decision limits (as proposed by Heitzman[9]) in the bovine already rely on separation of sex, age and gestation status as summarised in Table 2.1.

However, the use of hormones and other growth promoters is still legal in North America and other countries. This difference of opinion between Europe and other countries is the subject of an ongoing trade dispute moderated by the World Trade Organization (WTO).[10] The addition of growth hormones has resulted in 10–15% increases in daily gains, similar improvements in feed conversion efficiency (FCE) and improvement of carcass quality (increased lean/fat ratio). Thus, there has been a substantial reduction in the amount of energy required per unit weight of protein produced, and the economic implications of this have been great.[11] It can be difficult to prove that natural hormones have been administered to cattle if sufficient time is allowed between administration and monitoring and/or slaughter.

Hormones can be defined as natural (endogenous) or synthetic (exogenous), and can have anabolic or catabolic actions. These words are defined as follows:

1. Endogenous hormones are hormones naturally present in an animal and include (among others) oestradiol, progesterone and testosterone as well as their precursors and metabolites. Illegal administration of these compounds can give rise to elevated levels, but since certain physiological situations can also give rise to elevated levels, this makes proof of

Table 2.1 EEC decision limits for testosterone in bovine plasma (as proposed by Heitzmann[8])

Age/sex of animal	*EEC decision limit in plasma (ng ml^{-1})*	
	Oestradiol	*Testosterone*
Female (non-pregnant)	0.04	0.5
Male (< 6 months)	0.04	10
Male (> 6 months)	0.04	30

adulteration difficult. The Food and Drug Administration (FDA) has concluded "that no additional physiological effect will occur in individuals chronically ingesting animal tissues that contain an increase of endogenous sex steroids from exogenous sources equal to 1% or less of the amount in micrograms produced by daily synthesis in the segment of the population with the lowest daily production (pre-pubescent boys for oestradiol and pre-pubescent girls for testosterone)".[12]

2. Exogenous hormones are synthetic compounds which mimic the behaviour of the natural hormones. The most widely used of these are trenbolone, melengesterol acetate and zeranol, which is structurally related to the mycotoxin zearalenone.
3. Anabolic hormones promote muscle growth.
4. Catabolic hormones cause tissue breakdown and other metabolic effects.

There are two accepted practices for the monitoring of hormones in cattle:

1. The analysis of plasma, urine and faeces before or after slaughter. The detection and confirmation of steroids in a complex matrix such as urine are very difficult due to the wide variety of steroids, their metabolites and conjugates at a wide range of concentrations,[13] as well as the identical metabolites of natural steroids from any administered product.[14] Both steroids and their metabolites can be found in the free and conjugated forms.[11] Depending on the age and reproductive state of the animals, blood plasma concentrations can vary by an order of 1000, with oestrogens generally showing the lowest levels.[11] Hair has also been used as an indicator of hormone concentrations.[15]
2. The analysis of muscle, fat or organs of animals after slaughter. The indication of illegal injection falls under this latter category and meat from that animal would be classed as unfit for human consumption.

The task of detecting the abuse of endogenous hormones is problematic for many reasons. The most significant problem arises from the fact that when they are shown to occur naturally within a particular type of animal, a simple qualitative demonstration of their presence does not usually prove abuse. Some type of quantitative threshold concentration or ratio to another endogenous substance is therefore usually required in order to confirm abuse. Where particular steroids are believed not to be endogenous in a particular type of animal at a certain limit of detection, who is to say that, as analytical sensitivities improve, it won't be discovered that they are endogenous at a lower concentration?

Various analytical methods have been employed in the past to identify and quantify endogenous steroids and their metabolites and precursors, but their effectiveness and the harmonisation of their application in different countries and situations is questionable. Van Ginkel *et al.*[16] for example highlight the wide range of different analytical methods and thresholds that have in the past been applied in different EU countries. Since there is no comprehensive published

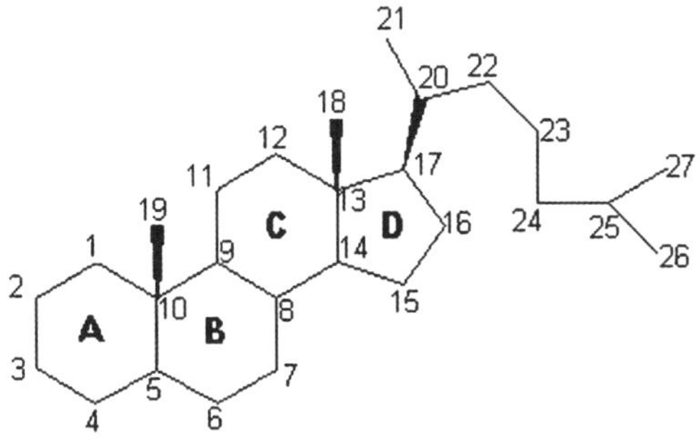

Figure 2.2 Steroid structure and numbering.

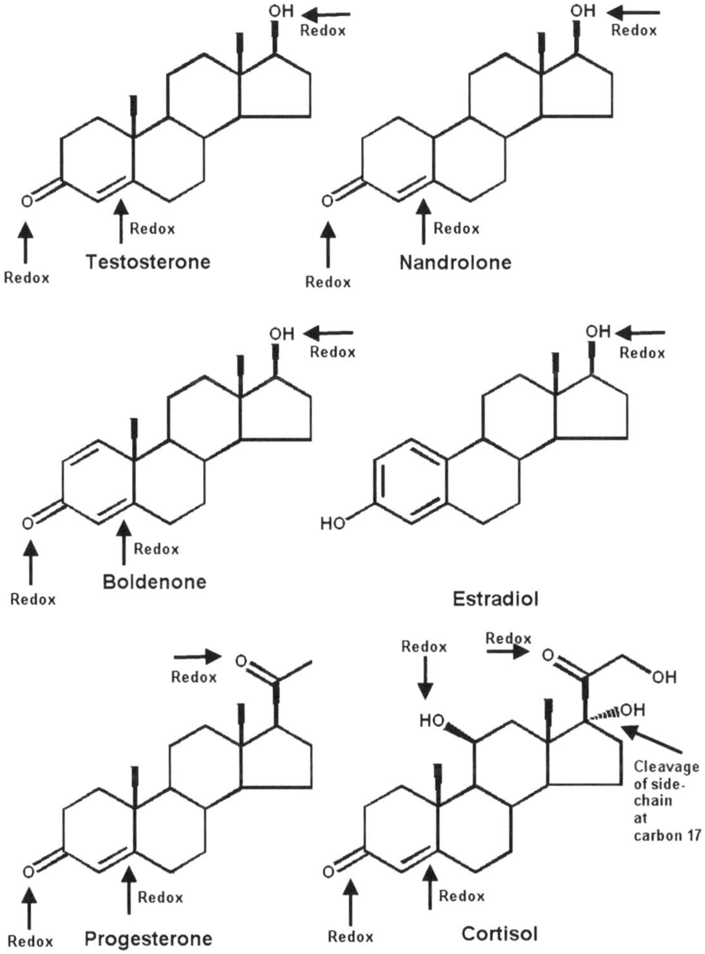

Figure 2.3 Major sites of endogenous steroid catabolism that are common to all species (highlighted by arrows).

Metabolism of testosterone, progesterone and 17β-oestradiol follows the steps outlined above, leading to rapid biological deactivation, producing the 17α-epimers for 17β-oestradiol and testosterone in cattle and sheep.[20] The same steps are not seen in pigs as they do not have the enzyme (possibly 17α-hydroxysteroid dehydrogenase[21]) to convert 17β-steroids to 17α-steroids.[22] Testosterone is efficiently reduced at the 4,5-double bond and at the 3-keto group by the enzymes 5α-reductase and 3 α/β-hydroxysteroid dehydrogenase respectively and is oxidised at the 17β-hydroxy group by 17-hydroxysteroid dehydrogenase. The very low concentrations of 17β-oestradiol in peripheral blood plasma suggest a rapid first-pass effect and/or an inactivation during passage of the intestinal wall.[11]

Different effects are produced by organ-specific patterns of steroid metabolising enzymes or by interaction with receptors of other steroid classes so that androgens can be aromatised to oestrogens in peripheral tissues or androgen metabolites can interact with oestrogen receptors.[3] The activity of steroid hormones is also regulated by the proportion of free circulating hormones and hormones which are bound to steroid hormone-binding proteins.[3] Metabolism leads to a rapid inactivation of many steroids (first-pass effect) and hence there is very little oral activity.[11]

Metzler makes some salient points in his recommendations for detection of natural hormones in urine:[23]

1. Metabolites may be present in greater quantities than the parent compound.
2. Metabolites may be more easily accessible and/or more suitable for analysis than the parent compound.
3. Metabolites determined in addition to the parent compound facilitate the interpretation and increase the reliability of analysis.
4. Metabolites may allow the detection of the exogenous administration of an anabolic compound which is also formed endogenously.

2.1.3 Conjugation

Phase II metabolism involves conjugation reactions. There are two major routes: sulfation and glucuronidation, both of which are affected by phase I reactions such as hydroxylation and reduction.[24] Biological glucuronidation and sulfation of steroids occurs by the transfer of glucuronic acid and sulfate moieties from uridine diphosphoglucuronic acid (UDP-glucuronic acid) and 3-phosphoadenosine-5-phosphosulfate (PAPS) respectively.[24] 17β-hydroxy-17α-methyl steroids are more likely to form sulfates than glucuronides due to steric hindrance of UDP-glucuronic acid at the C17 position.[24] The sulfate moiety in PAPS is much less hindered and relatively vulnerable to nucleophilic attack by the tertiary 17β-hydroxy group of these steroids.[24]

When substantial amounts of steroids are used, sulfation may become an important complementary phase II reaction that accelerates the elimination of the parent steroid and its metabolites (in humans).[24] Parental steroids and deactivated metabolites are eliminated with urine and bile following conjugation and, for cattle, it has been shown that 60–90% of the sex steroids are eliminated with bile *via* the faecal route.[25]

In order to assess more accurately the total amount of "free" steroid present in a sample, hydrolysis is often used prior to analysis. Hydrolysis can be performed before or after preliminary extraction or group separation and even then can be performed by a variety of methods. *Helix pomatia* digestive juice is the most often applied enzymatic form of deconjugation[26] and this method affords hydrolysis of glucuronic acid conjugates and aryl sulfates at optimum pH, but is known to contain hydroxylase and oxidoreductase enzyme activity that can artifactually oxidise or reduce some steroids.[27] Another preparation that is frequently used is the β-glucuronidase enzyme from *E. coli*, which as its name suggests cleaves glucuronic acid conjugates but not sulfate conjugates. In a two-fraction extraction, glucuronic acid conjugates may be cleaved by the *E. coli* enzyme, while sulfate conjugates can be cleaved by acid solvolysis. Another alternative is to cleave both types of conjugates simultaneously using methanolysis (methanol and acetyl chloride), but this can lead to a more complex mixture of components retained within extracts.[28] The use of a number of different hydrolysis (or no hydrolysis at all) steps in the literature, all with varying capacities to deconjugate steroids, is another factor that can potentially lead to variation in the reported concentrations.

When surveying the ranges of mean, minimum and maximum values among the published studies (over 1000 papers for all species concerned), approximate rank orders of absolute concentrations for different steroids in different matrices for each of the species can be constructed and are given in each of the following sections. It must be stressed that some positions within these ranks may be caused by biases in the amount of information reported for each steroid in different matrices. Due to the multiple variables that influence reported steroid concentrations, some of which were described in the above sections, it is not possible to provide a simple summary table of this data. The following sections therefore discuss the range of different steroid concentrations in the context of the different factors that lead to this variation *i.e.* species, sex, age, castration and pregnancy status, medication, diet, stress, housing conditions and breed, *etc.*

While most concentrations reported are in $ng\,ml^{-1}$ or $ng\,g^{-1}$, for the oestradiols they are generally reported in $pg\,ml^{-1}$, reflecting the much lower concentrations seen in animals and some testosterone concentrations are in mg^{-1}, reflecting the much higher endogenous concentrations seen in some animals. This review will focus on endogenous hormones.

2.2 Analytical and Physiological Considerations Regarding Comparisons of Steroid Concentrations within and between Different Species

A basic understanding of the analytical and physiological context of natural steroids is assumed in this review. Nevertheless, some information of specific relevance is given below and the biosynthetic and catabolic pathways of the major natural steroids are summarised in Figures 2.1 and 2.3 respectively. Further background information on general analytical aspects can be found in Makin *et al.*[29] and Stolker *et al.*,[8] whilst further physiological information can be found in Mason *et al.*[30] and Hadley and Levine.[31]

Although the background given here is separated into analytical and physiological factors, there are areas of overlap between the two. A critical theme that will become apparent is that even though it may sometimes be desirable to take into account a particular variable for analysis, the lack of reporting of this information (at least in a standard format) often means that rigorous quantitative comparisons are often not possible. It was also necessary to limit the number of analytical parameters chosen for study. The remaining parameters considered were chosen by consideration of a combination of their impact on any results as well as the frequency and reliability of their reporting.

2.2.1 Analytical Factors

Although most published methods rely on direct identification and/or quantification of analytes, indirect biomarker approaches have also been investigated. For example, the histological screening for illegal administration of growth-promoting agents in veal calves has been described.[32] This technique showed some potential for using the effects of androgens and oestrogens on male prostate or female clitoris/Bartholin gland as a biomarker of abuse, but has not been widely applied. For the purposes of the current review, studies were limited to direct detection/quantification using such techniques as immunoassay (IA), high performance liquid chromatography with ultraviolet detection (HPLC-UV), liquid or gas chromatography coupled to mass spectrometry (LC- and GC-MS respectively) and combustion isotope ratio mass spectrometry (GC-C-IRMS).

When comparing data between studies, it becomes apparent that while "true" differences between data points and populations do exist, variation can also be caused by relative biases in sampling designs or the type of analysis used. In many cases, comparison of data is further complicated by the reporting of different types of information. For example, limit of detection (LOD), decision limit (CCα) or detection capability (CCβ) are often not reported. Some examples of analytical aspects that can lead to variation within the data are given below:

2.2.1.1 Qualitative, Semi-quantitative and Fully Quantitative Data

While the ultimate aim of this review was to consider fully quantitative concentrations of natural steroids, it was also recognised that a number of useful studies only reported data in a qualitative or semi-quantitative fashion. The results of these analyses were of use in certain situations and are described. Where qualitative or semi-quantitative data are analysed, this will be highlighted and any assumptions stated. Even within fully quantitative data, a number of factors (as exemplified in the rest of this section) can lead to variability in the data set, so it is important to understand that there is a degree of uncertainty attached. In most studies, insufficient validation data were available to fully assess the degree of certainty of the results.

2.2.1.2 Type of Calibration Line Used for Quantification

When dealing with endogenous substances, quantification can sometimes be complicated by the difficulty of finding a true blank matrix. If a blank matrix of the same type as the study samples is not available then one can either use standard addition, where known amounts of steroids are added "on top" of the existing concentrations present, or alternatively a surrogate matrix can be used. If using a surrogate blank matrix, then appropriate measures need to be taken to ensure the chosen matrix behaves in a similar way to the actual sample matrix so that it can account for any variation in the analytical procedure. Neither of these measures is ideal, with the result that concentrations of steroid quantified using different calibration line approaches can lead to different reported values for the same data set due to differential matrix effects or recovery of analyte. In many of the reports reviewed, the calibration range applied was not explicitly given, making it difficult to evaluate whether individual results fell within a linear range. Due to the need to limit the number of factors that were being taken into account in this review, adjustment of analytical data for recovery and matrix effects was not attempted. In any case, many of the aforementioned parameters were not always reported by authors.

2.2.1.3 Limit of Detection (LOD) or CCα/β

For some steroids in certain physiological situations, a large number of reported concentrations are "not detectable" (ND). Unfortunately, these concentrations generally lie between the LOD and Limit of Quantification (LOQ) and therefore cannot be reliably quantified. This causes problems because these values have to be effectively treated as zero from a statistical point of view. The European Commission has recently introduced the concept of CCα and CCβ for regulatory purposes.[7] While all methods developed in Europe, or by their trading partners, now report values for CCα/β, the LOD is reported by the majority of other method developers. The majority of published methods do not differentiate between the LOD for the instrument used in the measurement and the LOD for the method itself.

2.2.1.4 Sample Collection and Subsequent Preparation Technique

Prior to analysis, most techniques require some degree of sample preparation, usually involving extraction of the analytes of interest from unwanted or interfering matrix components. The treatment of the sample once taken from the animal can influence the analytical results in several ways, all of which highlight the need to preserve samples appropriately and to take into account any artifactual processes occurring prior to analysis. For example, it is known that a number of meat-producing species, *e.g.* bovine, ovine and equine, have a significant capability to convert 17β-hydroxy functions or ketones into 17α-hydroxy compounds, and also that the porcine lacks the enzyme to perform this action. In the bovine this activity is known to be present in plasma,[33] but the addition of methanol to the matrix can limit the conversion reaction.

Enzyme hydrolysis is often used to break down the conjugates formed. This is discussed in the metabolism/conjugation section above (Section 2.1.3). A mixture of purification/concentration approaches were identified in the literature including solid phase extraction, liquid-liquid extraction, protein precipitation, immunoaffinity column chromatography, supercritical fluid extraction, accelerated solvent extraction (ASE) and some very elaborate, but often effective, multi-step HPLC fractionation processes. Results using these methods are generally not compared in this review, unless there was specific relevance to a result.

2.2.1.5 Type of Analytical Method Used

A major factor leading to variation between reported values lies in the type of method used. These included (in approximate descending order of reported use) immunoassay (IA), gas chromatography-mass spectrometry (GC-MS), liquid chromatography-mass spectrometry (LC-MS), high performance liquid chromatography-ultraviolet detection (HPLC-UV) and thin layer chromatography-ultraviolet or -fluorescence detection (TLC-UV/FL). Immunoassay and mass spectrometry techniques generally afford higher sensitivity over HPLC-UV or TLC-UV/FL and are also generally more specific. Mass spectrometry is considered to offer more selectivity than IA, predominantly due to variable extents of cross-reactivity of steroids against the IA antibody, although the impact of any cross-reactivity can be reduced by performing HPLC separation of sample extracts prior to analysis. Although more selective, mass spectrometry is still subject to matrix effects (LC-MS generally more so than GC-MS) and this can also affect the results. Ion suppression is also a problem with MS detectors, but this can be overcome to some extent by using matrix-matched standards. As a general rule, it has been observed that IA tends to overestimate oestrogen levels at low concentrations while underestimating them at high concentrations.[34]

2.2.1.6 Statistical Analyses Used Within the Studies Reviewed

Within residue and sports steroid testing, data sets are typically latitudinal in nature, encompassing both low as well as high values, rather than basing statistical analysis (*i.e.* threshold setting) on maximal steroid concentrations. This has the advantage of taking into account the whole range of possible natural variations, but does lead to a wider variance within the data compared with a maximal value approach. In the studies reviewed here a large number of parametric and non-parametric approaches have been reported. In this respect, it is important to highlight a major difference between a statistical method being able to discriminate a control from a steroid treated population (*i.e.* a T-test result) and a statistical method that allows a workable threshold to be calculated (*i.e.* allowing a degree of certainty that at a particular threshold a false positive will not occur). There can be a significant amount of overlap in individual steroid concentrations from control and treated steroid populations that can be discerned using a T-test, but this does not necessarily mean there are significant enough differences to allow a realistic threshold to be fixed.

2.2.2 Physiological Factors

Physiological factors play a very important role in the variation of observed concentrations. These include inter-individual and inter-species differences, the route of biosynthesis of steroids, pregnancy and pseudo-pregnancy, oestrous cycle and synchronisation, route of excretion, hydration status and other variables.

The peak ovarian cycle plasma/serum concentrations of oestradiol and progesterone in mammalian species have been studied[35] and it was found that oestradiol concentrations spanned approximately four orders of magnitude while those for progesterone spanned three orders of magnitude. Oestradiol concentrations were on average two orders of magnitude lower than progesterone concentrations and there were significant differences between the two different animals. Maximum oestradiol concentrations were more variable in artiodactyls and primates than in carnivores. Absolute oestradiol concentrations were not correlated with dietary niche, but the progesterone to oestradiol ratio was lower in artiodactyls and primates compared with carnivores. Although relating to oestrogens and progestagens rather than androgens (a comparable study for androgens could not be found by the authors), this study highlights the significant differences in steroid concentrations between species and identifies the need to obtain endogenous population data for hormones in each species before detection strategies are devised.

2.2.2.1 4- vs. 5-ene Pathways

Apart from the metabolic pathways described in Section 1.2 and the absolute differences in oestradiol and progesterone described above, the production of steroids using the 4- and 5-ene pathways is species dependant and can be traced back to differences in the substrate requirements of the CYP17 enzyme.[30] This means that some species, such as the cervine,[36] produce more steroid precursors with a 4-ene group (*e.g.* 4-androstenedione) whereas others, including the bovine, porcine, ovine and equine species, produce more with a 5-ene group (*e.g.* DHEA [dehydroepiandrosterone]).

2.2.2.2 Pregnancy and Pseudo-pregnancy

It is well known that pregnancy can lead to extremely high concentrations of certain relevant steroids, so pregnant animals are usually excluded from threshold value calculations. However, a phenomenon termed pseudo-pregnancy (also known as phantom pregnancy or pseudo-cyesis) also exists, which in some species can lead to the physiological appearance of pregnancy (including raised steroid concentrations), but without an actual foetus being conceived.[37] The effect is certainly frequent in rodent and canine species, but some references to its occurrence in the porcine,[38] caprine[39] and ovine-caprine hybrids[40] were also obtained. No reports of pseudo-pregnancy in bovine species could be found in the published literature. The effect of pregnancy on hormone concentrations is discussed later (Section 2.9).

2.2.2.3 Oestrous Synchronisation

The effects of oestrous synchronisation devices are not covered in this survey, but the subject has received a comprehensive review by Rathbone *et al.*[41]

2.2.2.4 Route of Excretion

Endogenous and artificially administered steroids are predominantly excreted from the body *via* the urine and faeces. Some steroids are preferentially excreted in faeces and others in urine, the route being species dependent. Consideration of whether urine, bile or faeces are the most suitable choices for a particular steroid/species combination depends on a number of factors (taken up later in this review), but their relative excretion in the form of recovered radioactivity in urine *versus* faeces is one method used to ascertain the best matrix to monitor.[42] Although an important factor, a predominance of radioactivity in one or other matrices does not always imply greater suitability for that matrix, as a smaller proportion of radioactivity present as one analyte may be more useful than a larger proportion of radioactivity present as many metabolites. Differences in the total volume of excreted material can influence resulting concentrations, while urine generally suffers fewer analytical matrix effects and residual *ex vivo* metabolism than faeces on the whole. Figure 2.4 exemplifies the range of different excretion patterns that have been observed for some steroids.

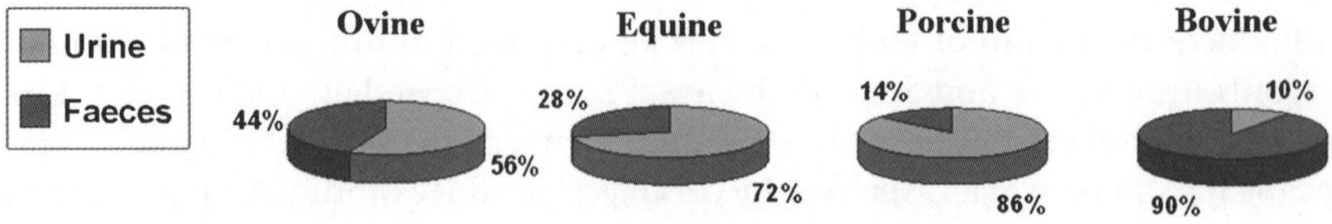

Figure 2.4 Percentage excretion of radioactivity in different waste products after intravenous infusion of testosterone into different species.

2.2.2.5 Hydration Status

The concentrations of steroids in some matrices, especially urine, can be affected by the hydration status of the animal.[43] One could predict that this might be a particularly important factor in countries that have experienced frequent droughts in recent years *e.g.* Australia. The correction of urinary steroid concentrations for hydration status, for example indirectly through measuring the specific gravity of the urine, therefore has potential to reduce the variation in population values. It can also be speculated that dehydration may act as a stressor on other parts of the body and thus independently affect minor metabolic pathways.

2.2.2.6 Other Variables

Many other physiological variables can affect the concentrations of steroids in different animals. Previously proposed regulatory thresholds for natural steroids in meat-producing species have taken into account at least the age and sex of the animal when constructing thresholds.[44,45] In the current review, some of the factors that were analysed include the steroid in question, matrix, age, sex, gestation and castration status, country of origin, season and time of day, disease, medication, housing conditions, diet and breed.

2.3 Endogenous Occurrence of Testosterone-related Compounds

Testosterone and related steroids such as epitestosterone, 4-androstenedione and DHEA are ubiquitous among male and female animals of all mammalian species, so differences among various groups and times are purely quantitative (see Table 2.2).

2.3.1 Bovine

By far the greatest amount of literature presented on the subject has involved the monitoring of testosterone concentrations in various matrices from the bovine. It has been noted that while several precursors including DHEA, androstenediol isomers and 4-androstenedione have occasionally been quantified in plasma/tissue, they were not monitored in excretory products. There may be value in monitoring perturbations of endogenous steroid feedback loops after exogenous steroid administration by monitoring levels of precursors as well as the steroids themselves.

Ranges of mean plasma/serum concentrations of testosterone and epitestosterone were found to be approximately ten-fold higher in intact mature males relative to females.[45] A study by Gerrard *et al.* compares the serum testosterone level of steers and bulls.[46] The authors note that while bull testosterone concentrations increase to a maximum of 4860 $ng\,ml^{-1}$ at 15 months,

Table 2.2 Summary of the endogenous occurrence of androgenic-anabolic steroids in mammalian meat-producing animal species. NMP = National Monitoring Programme, therefore not controlled with respect to steroid abuse

	Details of endogenous presence in different species		
Steroid	*Bovine*	*Ovine*	*Porcine*
Testosterone, oestradiol, progesterone and cortisol	Ubiquitous in males and females at varying concentrations	Ubiquitous in males and females at varying concentrations	Ubiquitous in males and females at varying concentrations
Nandrolone and related 19-nor androgens	Epinandrolone detectable during pregnancy. Most other studies find no nandrolone or related metabolites in males, but some find trace amounts in male and female urine i.e. following casualty (G. Kennedy – personal communication)	Epinandrolone detectable during pregnancy. One report of nandrolone and epinandrolone in both male and female urine in the UK, but population not controlled	Nandrolone and 19-norndrostenedione detected in urine and some other matrices of animals of all sexes (including intersex animals) at different concentrations
Boldenone and related 1-dehydro androgens	Boldenone and related compounds have been detected in urine and faeces, possibly secondary to their formation by gut bacteria. Some phase 1 + 2 metabolites only detected after boldenone admin	Insufficient data to draw any conclusions, although 2 of 961 urines from an Australian NMP contained low concentrations of epiboldenone	Boldenone detected in urine and some other matrices of boars, cryptorchids, gilts and barrows at different concentrations, but not above the LOD in sows or an intersex animal

steer concentrations remain constant. One significant finding was of a study that stated that plasma testosterone was exceptionally high for a very brief time during the late luteal phase of the normal female oestrous cycle, exceeding 1.8 ng ml^{-1}.[47] All other ranges of testosterone reported by this author were in-line with those of other studies so, if real, this phenomenon could have a serious, negative impact on the validity of the existing EU decision limit for females.

Several studies have assessed the effect of age on the plasma/serum concentrations of testosterone in males. Unfortunately, the different age ranges studied make meaningful comparisons of the data difficult. Bagu *et al.*[48] measured the mean male serum testosterone concentrations at 4, 20, 28 and 32 weeks. Their results, in common with other studies referenced in their paper, show a decrease in testosterone concentration between 20 and 32 weeks. However, another study on animals of the same age range revealed that the concentration of testosterone steadily increased from 15 to 23 weeks.[49] This study defined puberty as 17 weeks. A study on older male cattle, between 26 and 52 weeks of age, revealed concentrations increasing to a maximum of 8 ng ml^{-1} at 43 weeks and then decreasing at 52 weeks.[50] The same study also showed that 4-androstenedione concentrations decrease from 17 to 52 weeks. The most informative study on the effect of age was published by Arts *et al.*[51] While median male plasma testosterone concentrations increased from 0.8 ng ml^{-1} at 15 weeks to 1.3 ng ml^{-1} at 28 weeks, concentrations of epitestosterone dropped from 7.1 ng ml^{-1} at 15 weeks to 0.8 ng ml^{-1} at 28 weeks. As a result of the aforementioned testosterone and epitestosterone concentration changes with age, the epitestosterone:testosterone ratio fell significantly from 15 to 28 weeks of age. The testosterone:epitestosterone ratio has been found to be a good indicator of testosterone abuse in humans and horses (due to selective elevation of testosterone after testosterone doping), but Angeletti *et al.*[52] showed it to be of less use in the bovine, probably due to the relatively high 17α-hydroxylase enzyme activity.

The effect of age on female testosterone concentrations has not received so much attention. Mean female plasma testosterone concentrations immediately after birth have been reported at 0.075 ng ml^{-1} but decreased to means of 0.015 and 0.021 ng ml^{-1} between birth and puberty[53] where mean (SEM) age to puberty was 43.3 (1.3) weeks, with a range of 38–55 weeks. Median female plasma testosterone concentration was also monitored by Arts *et al.*,[51] who reported it as less than the LOD at both 15 and 28 weeks, while median plasma epitestosterone was less than the LOD at 15 weeks and then rose to 0.2 ng ml^{-1} at 28 weeks. Lactating cows of the Herens breed (artificially selected for fighting ability) had significantly higher ($P < 0.05$) median plasma testosterone concentrations compared to the Brune des Alpes breed, with values of 0.21 and 0.11 ng ml^{-1} respectively.[54] The biochemical observation was also borne out by secondary sexual characteristics, which were more prominent in the Herens breed.

A study by Moura *et al.*[50] showed that bulls suffering spermatic arrest had only slightly lower serum testosterone concentrations than healthy controls, but that serum concentration of 4-androstenedione in one diseased animal was $> 0.8\,\text{ng ml}^{-1}$ at 12 months, relative to a mean of $0.25\,\text{ng ml}^{-1}$ in healthy controls. No reports on the effect of other factors known to increase the androgen output in other species were found *i.e.* stress or congenital adrenal hyperplasia.

Ranges of mean urinary concentrations of testosterone and epitestosterone were found to be approximately three-fold higher in mature males compared to females. In veal calves testosterone was approximately three-fold higher in males compared to females (3.7 *vs.* $1.1\,\text{ng ml}^{-1}$), while for epitestosterone concentrations were $41\,\text{ng ml}^{-1}$ for males and $17\,\text{ng ml}^{-1}$ for females, all values measured at 28 weeks.[45] Very similar results were obtained in another study, where median male urinary testosterone concentrations were $1.0\,\text{ng ml}^{-1}$ at 15 weeks and $3.7\,\text{ng ml}^{-1}$ at 28 weeks. Epitestosterone concentrations did not change with age and at 15 and 28 weeks were 40 and $41\,\text{ng ml}^{-1}$ respectively.[51] Median female urinary testosterone concentrations were measured in the same study and were found to be less than the LOD at 15 weeks, increasing to $1.1\,\text{ng ml}^{-1}$ at 28 weeks.[51] Median female urinary epitestosterone at 15 and 28 weeks were 6 and $17\,\text{ng ml}^{-1}$ respectively. No urinary testosterone data were available for steers.

Testosterone mean muscle concentration ranges were found to be approximately ten-fold higher in mature males relative to females or steers, although epitestosterone was similar between steers and bulls. There were no epitestosterone data for females. Ranges of mean liver and kidney concentrations of testosterone were found to be approximately ten-fold higher in mature males relative to females. There were no data for androgens in steers or epitestosterone in any sex. Ranges of mean hair concentrations of testosterone were found to be approximately three-fold higher in mature males relative to females and steers and there were no data available for epitestosterone.

Despite extensive literature searches, there were insufficient data to compare the effect of age or sex on testosterone or related metabolites/precursors in faeces, liver, kidney, bile, muscle, hair or fat. Neither were any studies found that directly compared concentrations of testosterone or related precursors/metabolites in similar breeds under different housing conditions or in different countries. There were also no studies found on the effect of diet, time of day or season on testosterone or related precursor/metabolite concentrations in the bovine.

In most species long-term treatment with gonadotrophin-releasing hormone (GnRH) agonists, such as deslorelin, decrease luteinising hormone (LH) output (and therefore testosterone secretion) due to desensitisation of the pituitary gland.[55] However, Aspden *et al.*[56] reported that testosterone concentrations in mature bulls are increased following deslorelin administration, although another effect of this drug is that LH pulsatility is lost, leading to a flat LH secretion profile. On the other hand, Renaville *et al.*[49] showed that administration of GnRH to immature bulls between 70 and 50 days of age delayed puberty relative to controls with mean pubarche ages of 180 and 120 days respectively. No reports of the effects of other non-steroidal medications on androgen concentrations were found, but several types of medication in other species are known to increase or decrease concentrations, *e.g.* cytochrome P450 enzyme inducing and/or inhibiting drugs.

Table 2.2 summarises the occurrence of testosterone in the bovine while Table 2.3 summarises the major metabolites following exogenous administration.

From the data available, an approximate rank order for testosterone concentrations in bovine matrices can be suggested: hair > urine ~ fat ~ faeces ~ kidney > plasma > liver ~ muscle, and for the major testosterone metabolite epitestosterone as urine > faeces > plasma > muscle > hair. No data was available for epitestosterone in fat, liver or kidney. In terms of absolute values, testosterone and epitestosterone were present at similar concentrations in muscle and plasma, testosterone was at least a factor of ten higher in hair, while epitestosterone was a factor of ten higher in urine and faeces. There were no data for comparison of fat, liver or kidney. There was more variation among epitestosterone values relative to those for testosterone. The majority of plasma results that contributed to the aforementioned results do not use sample hydrolysis. However, Scippo *et al.*[44] showed that while the maximum testosterone concentrations found in bull plasma were 5752 and 965 $pg\,ml^{-1}$ for unconjugated and conjugated, respectively, the reverse was seen for epitestosterone with values of 974 and 1750 $pg\,ml^{-1}$ for unconjugated and conjugated, respectively. This could lead to artificially low reported concentrations of epitestosterone in plasma relative to testosterone.

2.3.2 Ovine

Scarth *et al.* 2009 found ranges of mean plasma/serum concentrations of testosterone to be between three- and 100-fold higher in rams relative to ewes (depending upon the season), while concentrations in wethers were similar to those found in ewes.[57] Concentrations of 4-androstenedione in the only report of concentrations in wethers were similar to the lowest mean concentrations reported in rams, but approximately 60-fold lower than the highest mean concentrations reported in rams. Plasma/serum concentrations of testosterone in late pregnant ewes were similar to those of non-pregnant animals, while pregnant ewe DHEA concentrations were similar to those in rams. 4-androstenedione concentrations in pregnant ewes were similar to the maximum mean concentrations reported in rams and approximately 60-fold higher than the only report of concentrations in wethers.

The effect of age on plasma/serum testosterone concentrations in rams has been reported. Fahmy *et al.*[58] monitored rams of the Romanov and the Booroola Merina x DLS breeds from the ages of 10 to 34 weeks. Testosterone concentrations increased with age in both breeds ($P < 0.01$) with mean concentrations at 10 weeks of 3.8 $ng\,ml^{-1}$ and 0.8 $ng\,ml^{-1}$ in Romanov and Boorola Merina x DLS breeds respectively. At 34 weeks these concentrations were 13 and 8.0 $ng\,ml^{-1}$ respectively. Concentrations were significantly higher ($P < 0.05$) in the Romanov breed at all ages other than 14 weeks, which

Table 2.3 Summary of the major catabolic pathways of endogenous androgenic-anabolic steroids in mammalian meat-producing animal species. The metabolism of progestagens and corticosteroids are not shown because data are lacking. Oestrone is a major metabolite of oestradiol in all species, while α-oestradiol is also a major metabolite in the ovine and bovine.

	Details of major phase 1 urinary metabolic products after administration to different species		
Steroid	*Bovine*	*Ovine*	*Porcine*
Testosterone	Epitestosterone, epietiocholanolone, 5β-androstane-3α,17β-diol + 5β-androstane-3β,17α-diol	Epitestosterone, androsterone, etiocholanolone + 5β-androstane-3α,17β-diol	Insufficient data to draw any conclusions
Nandrolone	Epinandrolone, 5β-estrane-3α,17β-diol, 5β-estrane-3α,17α-diol, 5α-estrane-3β,17α-diol	Insufficient data to draw any conclusions	19-norepiandrosterone, 19-noretiocholanolone + 5β-estrane-3α,17β-diol
Boldenone	Epiboldenone, 17α-hydroxy-5β-androst-1-en-3-one, 6β-hydroxyepiboldenone, 17β-hydroxy-5β-androst-1-en-3-one + 3α-hydroxy-5β-androst-1-en-17-one	Insufficient data to draw any conclusions	Insufficient data to draw any conclusions

corresponded to a temporary reduction in concentrations in the Romanov breed.

Serum concentrations of testosterone in four different breeds at six, eight and 12 months then at three years were studied by Langford *et al.*[59] Mean concentrations in the Canadian Arcott, Outaouais Arcott, Rideau Arcott and Finnish Landrace at six months were 3.3, 4.0, 3.5 and 3.9 $ng\,ml^{-1}$ respectively, at eight months were 3.1, 5.0, 5.5 and 8.2 $ng\,ml^{-1}$ respectively, and at 12 months were 3.5, 5.3, 5.4 and 7.8 $ng\,ml^{-1}$ respectively. Mean concentrations at three years were 8.0, 7.0, 5.7 and 7.0 $ng\,ml^{-1}$ respectively. Concentrations were significantly higher ($P < 0.05$) in Finnish relative to Canadian Arcott male lambs at eight and 12 months, but none of the other differences were significant ($P < 0.05$). The results at 34 weeks in the Fahmy study[58] can be compared to those of eight months in the second study as the animals are the same age. The concentration in Romanov rams is comparable to that of the Finnish Landrace while those in Booroloa Merina x DLS are significantly higher. This suggests that testosterone concentrations are indeed species as well as age dependant.

There were insufficient data to compare concentrations of testosterone or related precursors/metabolites between sexes, ages, gestation and castration statuses in any other matrix.

No reports on the effect of time of day or geographical factors were found, but two reports on the effect of diet and several reports on the effect of season and housing conditions were retrieved. The effect of transferring animals from a 12-hour light, 12-hour dark photoperiod to either short day photoperiod (8-hour light, 16-hour dark) or long day photoperiod (16-hour light, 8-hour dark) on serum testosterone concentrations in Suffolk-Hampshire rams was studied by Schanbacher *et al.*[60] Short day exposure resulted in increased testosterone concentrations, heavier testes, larger seminiferous tubules and greater sperm production relative to long day exposure rams. This effect of photoperiod has obvious implication for resulting testosterone concentrations and its dependence upon the type of artificial light. Borque *et al.*[61] studied the effect of season on the concentration of plasma testosterone in Manchego rams living in the northern hemisphere (Spain). Concentrations varied with the season, with mean peak values in the second week of September (end of summer/beginning of autumn) at 7.2 $ng\,ml^{-1}$, while lowest concentrations of 0.40 $ng\,ml^{-1}$ were observed in the second week of February (winter). The authors also commented that increasing testosterone concentrations were correlated with a decreasing photoperiod.

The effect of housing males with females on plasma testosterone concentrations in Texel and Suffolk rams showed that testosterone concentrations increased on introduction to ewes ($P < 0.05$), but the effect did not depend on whether ewes were in oestrous or not.[62] Mean basal concentrations in different groups varied from 5 to 10 $ng\,ml^{-1}$, while, after introduction to ewes, concentrations increased to between 8 and 15 $ng\,ml^{-1}$.

Parkinson *et al.*[63] studied the effect of intersex relative to concentrations in "normal" rams and ewes during the breeding season. The authors classified Freemartin ewes (XX/XY chimaeras) as either "male (MF)" type or "undifferentiated (UF)" depending upon the masculinisation of their genitalia. In one of the experiments (1b) mean basal testosterone concentrations in MF, UF, ewes and rams were 0.79, 0.29, 0.14 and 2.4 $ng\,ml^{-1}$ respectively. The testosterone concentrations in Freemartins were significantly higher ($P < 0.05$) than in ewes, but lower than in rams.

Table 2.2 summarises the occurrence of testosterone in the ovine while Table 2.3 summarises the major metabolites following exogenous administration.

2.3.3 Porcine

Scarth *et al.* 2009 found concentration ranges of mean plasma/serum testosterone in boars to be between 7 and 500-fold greater than those observed in barrows (castrated males), while those in gilts/sows were around five-fold lower than those in barrows.[57] There were insufficient data to compare the concentrations of any other steroids by sex as there were few reports of testosterone concentrations in barrows and gilts/sows.

The most extensive study of variation of steroid concentrations with age in boar plasma (Yorkshire breed) was carried out by Schwarzenberger *et al.* in 1993.[64] Before detailing the results, it is important to point out that the concentrations reported seem to be much higher than those reported by other authors. The authors commented that steroids were almost exclusively present in plasma as sulfate conjugates so, unless indicated, the results are for sulfated steroids. 4-androstenedione and DHEA-sulfate were measured directly, whereas other steroids were measured after extraction and solvolysis of separated fractions using acidified ethyl acetate (a standard approach used to cleave the sulfate conjugates,[28] thus inferring, but not proving, the status of most conjugates. Their results are summarised in Table 2.4. Overall, sulfate conjugates were (or were inferred to be) the most predominant steroids measured in plasma (with the exception of 4-androstenedione, which was measured as free only). Concentrations of most steroids peaked first at one month after birth and then dropped until five months, before increasing again at six to eight months (with some dropping slightly after seven months). The rank order of absolute

Table 2.4 Summary of plasma steroid levels in boars at birth and in the subsequent months. All concentrations are $ng\,ml^{-1}$. Taken from Schwarzenberger *et al.*[63]

	Month					
	Birth	*1*	*5*	*6*	*7*	*8*
Free testosterone	1	3	1	3	–	–
Testosterone sulfate	1	3	1	–	6	–
4-andorstenedione	6	24	6	18	–	9
DHEA-sulfate	3	80	15	80	–	60
5α-androstane-3β,17β-diol-sulfate	3	12	3–6	2	35	–

determined concentrations in plasma was DHEA-sulfate > 5α-androstane-3β,17β-diol-sulfate > 4-androstenedione > testosterone sulfate.

McCoard *et al.*[65] studied the concentrations of plasma testosterone in Meishan and White Composite foetal and neonatal boars. While foetal testosterone concentrations were not significantly different ($P < 0.05$) in the two breeds, with values ranging from around 1 to 2 ng ml^{-1}, they increased neonatally in both breeds, peaking at 14 days post-partum with concentrations of approximately 5.5 and 7 ng ml^{-1} in Meishan and White Composites respectively (significantly higher in White Composites at $P < 0.05$). Concentrations then dropped to around 4 ng ml^{-1} in both breeds by 25 days post-partum. The effect of seasonal variation on serum testosterone concentrations in Duroc *versus* Yorkshire boars living in South Korea (a northern hemisphere country) was reported by Park *et al.*[66] Mean testosterone concentrations in Duroc boars during spring, summer, autumn and winter were 3.1, 0.73, 1.3 and 1.4 ng ml^{-1} respectively, while in Yorkshire boars were 5.1, 2.6, 2.5 and 2.6 ng ml^{-1} respectively. The higher testosterone concentrations in spring occur at a time when photoperiod is increasing in this country, but peak photoperiod (during the summer) does not correlate with peak testosterone concentrations. The testosterone concentrations in Yorkshire boars were significantly higher ($P < 0.05$) than for Duroc boars at all stages of the year. Testosterone concentrations were higher in all breeds in spring compared with the rest of year ($P < 0.05$). It would appear that the effect of photoperiod is opposite in pigs and sheep. Walker *et al.*[67] measured plasma testosterone concentrations in Duroc boars that had or had not been subject to selection for high testosterone concentrations over ten generations. Mean testosterone concentrations in boars selected over ten generations were 49 ng ml^{-1}, which were significantly higher ($P < 0.01$) than controls with a mean testosterone concentration of 28 ng ml^{-1}, also much higher than the South Korean study.

No reports of the effect of pregnancy, diet, housing conditions or country of origin on testosterone concentrations were found.

Table 2.2 summarises the occurrence of testosterone in the porcine.

2.4 Endogenous Occurrence of Nandrolone-related Compounds

Nandrolone was once thought to be a solely synthetic steroid, but in the 1980s it was isolated as a natural hormone in the stallion[68] and boar.[69] Since then nandrolone-related compounds have also been detected in matrices originating in the bovine,[70] ovine,[71] caprine,[72] human[73] and cervine.[74] Non-phenolic C18 steroids (19-nor androgens) such as nandrolone are thought to be predominantly produced as by-products of the normal aromatisation process of androgens that produce the phenolic oestrogens. They are therefore most often detected in situations of high oestrogen output such as pregnancy,[75] although there are other contributory causes. These include consumption of contaminated dietary products,[76] increased physiological stress[77] and *in situ* formation in urine samples,[78] stored either in the laboratory or in the bladder. The formation of 19-norsteroids by demethylation of endogenous steroids in urine has been observed. An elevated temperature appears to increase this conversion, but the addition of metabolic inhibitors, such as EDTA, to the samples appears to stop the reaction to some extent.[78] The possibility that 19-nor androgens may arise as artifactual products of 19-carboxy compounds, as seen in the equine, also has to be considered because most methods to date have not taken this possibility into account.[79]

2.4.1 Bovine

A representative selection of studies reporting endogenous bovine nandrolone are summarised below.

The first report of the detection of nandrolone-related compounds was in 1993 when traces of 5α-estrane-3β,17α-diol were detected in urine from a pregnant control cow and 19-nor-etiocholanolone was detected in two control steers by GC-MS (the breed was not reported). Nandrolone and epinandrolone were not detected in the cow or steer urine and no C18 androgens were present in calf urine. The method was not quantitative and no LODs were reported.[80]

A multi-laboratory study on the natural occurrence of C18 androgens in the bovine showed that although no nandrolone could be detected at an LOD of 0.5 ng ml^{-1}, urine from pregnant cows (the breed was not reported) contained epinandrolone for up to four months pre-partum and two days post-partum when analysed by GC-MS.[81] Although most laboratories agreed when epinandrolone was or was not present in a sample (with some exceptions probably due to concentrations close to LODs), there was significant variation in the quantified concentrations when the compound was found (*e.g.* between 0.7 and 4.3 ng ml^{-1} for one sample).[81]

The natural occurrence of nandrolone and epinandrolone in the bile of pregnant Friesian cows was studied using GC-MS.[82] Nandrolone itself was not detected in any sample from all stages of pregnancy (LOD not reported), but again epinandrolone was detected from 120 days pre-partum onwards, when concentrations were approximately 1 ng ml^{-1}, rising to 37 ng ml^{-1} at parturition and then dropping to not detectable within one week post-partum. Cows carrying male foetuses had higher ($P < 0.001$) epinandrolone concentrations than those carrying female foetuses and cows had lower ($P < 0.001$) epinandrolone concentrations during their second pregnancy relative to their first. The same authors subsequently showed that the bile of steers and bulls derived from an untreated population did not contain nandrolone or epinandrolone above the instrumental LOD of 0.4 ng ml^{-1}. However, some bile samples from steers (but not from bulls) suspected of nandrolone abuse did contain epinandrolone.[83]

Table 2.2 summarises the occurrence of nandrolone in the bovine while Table 2.3 summarises the major metabolites following exogenous administration.

2.4.2 Ovine

Very few reports on the occurrence of endogenous concentrations of 19-nor-androgens in the urine of the ovine were found. In a 1991 study by Vandenbroeck *et al.*[70] the urine of 11 rams/lambs and 10 pregnant/non-pregnant ewes were positive for nandrolone at a mean concentration of 2.5 ng ml^{-1} when analysed by RIA, but the samples could not be confirmed positive by GC-MS (the LOD was not reported).

The urine of four ewes at different stages of pregnancy and the amniotic fluid of one ewe were analysed for the presence of epinandrolone.[74] Animals were sampled between 43 and zero days prior to parturition, some at multiple times and others only once. The amniotic fluid did not contain epinandrolone above the LOD of 1 ng ml^{-1} but urine was found to contain epinandrolone at concentrations ranging from below the LOD to above 2 ng ml^{-1}. There was no clear correlation between the stage of gestation and the concentration of epinandrolone determined. The ages, parity and foetal number of the animals were not reported.

Another study for nandrolone and epinandrolone in 30 pregnant and non-pregnant four-year-old French Vendenne ewes using a GC-MS method with a reported LOD of 0.2 ng ml^{-1}.[71] Nandrolone was not observed in any animals and epinandrolone was detected at concentrations below 0.5 ng ml^{-1} in pregnant animals at 120 to 39 days before parturition, with concentrations then increasing to 3.4 ng ml^{-1} at seven days before parturition. There was no correlation between the sex or number of foetuses with the epinandrolone concentrations found.

The above results contrast with those of Sterk *et al.*,[72] who could not detect nandrolone or epinandrolone in the urine of five Flevolander ewes during early or late stages of pregnancy using a GC-MS method with an LOD of 0.5 ng ml^{-1}. The ages, parity and foetal number of the animals were not reported.

It was observed in the UK that the incidence of nandrolone positives rose once an LC-MS screening method with an LOD of 0.5 ng ml^{-1} replaced an ELISA method with an LOD of 2 ng ml^{-1} in 2004, resulting in a study that reported on the use of LC-MS to assess whether nandrolone and epinandrolone were natural in a population of 130 male and female sheep in the UK.[84] Although the population could not be guaranteed "clean", the authors report that no evidence of steroids abuse was found on any of the farms tested. The authors made the following observation:

"Nortestosterone seems endemic in British sheep; primarily as the 17α-isomer, but also with some 17β- present. There does not seem to be much correlation with age or sex of the animal, although the majority of the population tested was 6-12 months and some of the other categories contained very few samples (e.g. all 5 males of over 12-months contained the 17α-isomer at 0.4ng ml^{-1} or greater). In light of this, it seems unwise to extrapolate the '17α- in male animals indicates abuse' rules from cattle to sheep. An exercise to test the urine of a controlled population will validate these conclusions, and demonstrate any link to other physiological factors such as breed or feedings regime."

Table 2.2 summarises the occurrence of nandrolone in the ovine.

2.4.3 Porcine

There has been some interest in the occurrence of nandrolone in the porcine, since relatively high concentrations of nanodrolone have been detected in urine. The most comprehensive analysis of nandrolone- and boldenone-related compounds in the porcine to date was conducted by Poelmans *et al.*[85] Samples of muscle, liver, kidney, testicles and urine from boars, cryptorchids, barrows, gilts and sows from France, the Netherlands, Belgium and the USA were analysed for boldenone, nandrolone and 19-nor-4-androstenedione using GC-MS and LC-MS (with n = between 5 and 14 for the different sexes). The results of this study are summarised in Table 2.5. Limits of detection were not reported. Since barrows are castrated males, the major source of steroid production in this animal is likely to be the adrenal gland.

Van Cruchten *et al.*[86] reported the presence of 19-nor-androgens in what initially appeared to be a female pig. However, close inspection determined that the animal was in fact a hermaphrodite (intersex animal) due to the presence of both a left ovary and a right undescended testicle (with functioning leydig cells, but no spermatozoa). Samples of urine, fat, faeces, liver, kidney, muscle and testes were analysed for nandrolone and 19-nor-4-androstenedione using GC-MS for fat, urine and faeces and LC-MS for the remaining matrices. Concentrations of nandrolone in fat, urine, faeces, liver, kidney, muscle and testes were 0.3 ng ml^{-1}, 27 ng ml^{-1}, not detectable, not detectable, 1.6 ng ml^{-1}, not detectable and 5.3 ng ml^{-1} respectively, while concentrations of 19-nor-4-androstenedione in these matrices were not detectable, 0.5 ng ml^{-1}, not detectable, not detectable, 1.1 ng ml^{-1}, not detectable and 0.9 ng ml^{-1} respectively. The authors proposed visual inspection of pig external sexual organs in female pigs suspected of nandrolone abuse at slaughter in order to discern false positives due to intersex. Similar results were observed in the study by Poelmans *et al.*[85]

A metabolism study of nandrolone in Gottinger x Vietnamese mini-pigs before and after injection of nandrolone laurate showed that nandrolone was detected in boar urine, but not barrow or sow urine, prior to administration and in all animals after administration.[87] No 17-keto, A-ring reduced compounds were detected in any of the animals before nandrolone administration, but after administration 19-noretiocholanolone and 19-norepiandrosterone were detected in all animals. 19-norandrosterone was detected in barrows after nandrolone administration, but not in the boar or sow, while 19-norepietiocholanolone was not detected after nandrolone administration in any animal. No estranediols were detected before nandrolone administration in any animal. Neither 5α-estrane-3α,17β-diol or 5α-estrane-3β,17α-diol were detected after nandrolone administration, but two other estranediols, proposed to be 5β-estrane-3α,17β-diol and 5α-estrane-3β,17β-diol, were found in some of the animals.

Roig *et al.*[88] reported the results of a quantitative metabolism study following IM nandrolone laurate injection to a ten-week-old boar. Only post-administration urine results were reported. Following the administration, nandrolone was predominantly conjugated with sulfate with peak concentrations of approximately 80 ng ml^{-1} at day one dropping to around 10 ng ml^{-1} by day four.

Table 2.5 Concentrations of nandrolone, 19-nor-4-androstenedione and boldenone quantified by Poelmans *et al.*[100] in different matrices from different sex porcine animals (the number of animals are given in brackets. < refers to concentrations below the LOD)

Nandrolone ($ng\,g^{-1}$)	*Gilt*	*Sow*	*Boar*	*Barrow*	*Cryptorchid*
Meat	< (11)	0.4–0.5 (11)	0.7–13.4 (11)	0.7–11.8 (11)	0.1–2.4 (11)
Liver	0.1–0.9 (11)	< (11)	1–63 (11)	< (11)	0.2–12.3 (14)
Kidney	0.2–0.5 (11)	0.2–1.5 (11)	2.5–232 (11)	0.1 (10)	1.3–78 (14)
Urine ($ng\,ml^{-1}$)	1.3–2.8 (11)	1.3–1.9 (9)	51–344 (11)	0.5–16.3 (11)	8.6–343 (14)
Testes	–	–	24–144 (5)	–	2.2–101 (11)
19-nor-4-androstenedione ($ng\,g^{-1}$)	Gilt	Sow	Boar	Barrow	Cryptorchid
Meat	< (11)	0.04–0.07 (11)	0.1–5.5 (11)	0.05–0.8 (11)	0.04–0.4 (11)
Liver	0.3–1.3 (11)	3.1–8.3 (11)	0.1–24 (11)	1.9–16 (11)	0.4–3.5 (14)
Kidney	8.3–25 (11)	2.7–18 (11)	2.3–535 (11)	0.1–15 (10)	0.2–159 (14)
Urine ($ng\,ml^{-1}$)	1.8–17 (11)	0.9–18 (8)	5–109 (11)	1.1–16 (11)	9.9–103 (14)
Testes	–	–	6.2–110 (5)	–	1.3–25 (11)
Boldenone ($ng\,g^{-1}$)	Gilt	Sow	Boar	Barrow	Cryptorchid
Meat	< (11)	< (11)	0.5–2.5 (11)	< (11)	0.7 (11)
Liver	< (11)	< (11)	1.3–4.9 (11)	< (11)	0.5–2.3 (11)
Kidney	< (11)	< (11)	0.8–9.2 (11)	< (10)	0.3–8.1 (14)
Urine ($ng\,ml^{-1}$)	0.5–0.6 (11)	< (11)	5.1–120.5 (11)	1.1 (11)	0.9–57.6 (14)
Testes	–	–	2.1–16 (5)	–	0.6–15.1 (11)

Peak concentrations of different analytes in the glucuronic acid conjugate fraction were generally lower than 20 ng ml^{-1} while in the free fraction, norepiandrosterone, noretiocholanolone and 5β-estrane-3α,17β-diol were most abundant with peak concentrations of 40, 60 and 100 ng ml^{-1} respectively.

There has been some discussion about whether 19-nor-androgens in the porcine are excreted as 19-nor compounds or whether they are artefacts of sample preparation produced from an initial 19-carboxy metabolite, analogous to the situation in the male horse.[79] Several lines of evidence support this hypothesis, including the observation that no A-ring reduced metabolites are found in untreated boar urine.[87] Such compounds would be expected to occur as metabolic products if free nandrolone and/or 19-nor-4-androstenedione were present *in vivo*. Their absence suggests that it is either a directly secreted conjugate or a carboxylic acid metabolite present in urine, as these would be unlikely to undergo A-ring metabolism, but would be cleaved to form nandrolone and 19-nor-4-androstenedione during typical sample preparation procedures. The 19-carboxylic metabolite of 4-androstenedione has also previously been identified in porcine granulosa cells.[89] Lastly, an unusually high ratio of 19-nor-androgens in boar urine, relative to faeces, has been observed.[90] This adds further weight to the theory of 19-nor-androgens in the porcine actually being the 19-carboxylic acid, as these acidic compounds could be substrates for organic anion transport proteins in the liver/gut that actively transport compounds from one area to another and maintain large concentration gradients (something that is less likely to occur for neutral steroids).

If the compounds in porcine tissues that are currently thought to be 19-norandrogens are proven to be predominantly 19-carboxylic acids, similar to the situation in the equine, then this could have major implications for nandrolone residue screening methods in the porcine. For example, Roig *et al.*[88] showed that the major metabolites of nandrolone after exogenous administration to the boar are excreted in the free sulfate and glucuronic acid conjugate fraction, while Debruyckere *et al.*[87] showed that A-ring reduced metabolites are only detectable after nandrolone administration. It may therefore be feasible to use a threshold of nandrolone sulfate or an A-ring reduced metabolite as an indicator of nandrolone abuse in this species.

Table 2.2 summarises the occurrence of nandrolone in the porcine while Table 2.3 summarises the major metabolites following exogenous administration.

2.5 Endogenous Occurrence of Boldenone-related Compounds

As in the case of nandrolone, boldenone and other 1-dehydro steroids were once thought not to be endogenous. Since the 1990s however, boldenone-related compounds have been detected in different matrices from several species including microbes,[91] maggots,[92] crustaceans,[93] rats,[94] pigs,[95] horses[96] and cattle (veal calves).[97] The most likely origin of 1-dehydro compounds in the bovine is through faecal conversion of precursors such as phytosterols or other steroids by gut microbes.[98] Extra-enteral production of 1-dehydro steroids within the body (*i.e.* testes) has not been demonstrated in the bovine,[22] but boldenone has been identified in porcine testes.[95] A very comprehensive review of the presence and metabolism of boldenone in various animal species was published by De Brabander *et al.* in 2004.[22]

2.5.1 Bovine

The endogenous occurrence of 1-dehydro steroids in the bovine has been the subject of some debate as they appear to be present in some animal populations but not others. This is highlighted in the following two studies.

Urine samples were analysed for boldenone, epiboldenone and androsta-1,4-diene-3,17-dione by LC-MS from 25 untreated animals and it was found that boldenone (LOQ = 0.2 ng ml^{-1}), epiboldenone (LOQ = 0.5 ng ml^{-1}) and androsta-1,4-diene-3,17-dione (LOQ = 0.2 ng ml^{-1}) were not detected above the LOQ in any of the blank urine samples from the untreated animals (Draisci *et al.*).[99]

Bovine faeces is known to be capable of producing boldenone and other 1-dehydro steroids as metabolites from some steroidal precursors *ex vivo*[98] as outlined in a report by Pompa *et al.*,[98] who studied the concentrations of boldenone, epiboldenone, androsta-1,4-diene-3,17-dione, testosterone and epitestosterone in the urine, skin swabs and faeces of Friesian calves and also assessed the effect of drying the faeces on the resulting faecal steroid concentrations. In urine, LODs for all steroids were 0.1 ng ml^{-1} and in faeces they were 0.5 ng g^{-1} (based on S:N > 3 : 1). Boldenone, epiboldenone and androsta-1,4-diene-3,17-dione were not detected in any of the urine samples from 10 calves, while boldenone was detected in rectal faeces from all the calves at concentrations ranging from 28 to 89 ng g^{-1}. Epiboldenone in rectal faeces was not detectable in six calves and ranged from 2.6 to 5.9 ng g^{-1} in the other four animals. Androsta-1,4-diene-3,17-dione in rectal faeces was not detected in nine calves, while one calf had a concentration of 21 ng g^{-1}. Results from skin-scraped rectal faeces, faeces taken from the stall floor and faeces allowed to dry for up to 13 days in the cowshed showed that the concentrations of all steroids increased significantly (but variably) over time. This is especially true of epiboldenone and androsta-1,4-diene-3,17-dione, which by day 13 of storage are present in high concentrations, while boldenone was reduced to not detectable by day 13. This study exemplifies the need for avoiding faecal contamination of urine during sampling and to ensure swift storage and analysis of any samples taken in order to avoid boldenone false positives.[22]

Table 2.2 summarises the occurrence of boldenone in the bovine while Table 2.3 summarises the major metabolites following exogenous administration.

2.5.1.1 EU Recommendations Regarding Boldenone Testing in the Bovine

The outcome of an experts meeting on the control of boldenone in veal calves in September 2003[100] recommended the following:

"On the basis of the scientific information available, the experts of the Member States agreed that the presence of boldenone conjugates at any levels in urine from veal calves is proof of illegal treatment. In order for positive results for boldenone to be used as evidence of illegal treatment, the following must be fulfilled:

- *That sampling of urine must be done without faecal contamination of the samples. The samples should be frozen as soon as possible after collection in order to avoid hydrolysis of the conjugates.*
- *Analytical results related to boldenone residues (boldenone or epiboldenone) must always be specified as free or conjugated forms, with the explicit identification of the animal species, including breed, gender and age of the animal.*

There is sufficient scientific knowledge to conclude that the presence of epiboldenone in urine and faeces of bovine animals can come from other sources than illegal treatment. A number of explanations are currently being investigated by the scientific community. If only epiboldenone is found and if the levels are above 2 ng g^{-1} in urine of veal calves, additional investigations would need to be carried out before concluding an illegal use of boldenone.

An MRPL for the analytical methods for the detection of boldenone and epiboldenone in urine of veal calves should be set at 1 ng g^{-1}. Further studies of appropriate marker metabolites of boldenone are encouraged. The member states should transmit existing and future data to the CRL in Bilthoven. This position could be amended in the light of additional data from ongoing and future research."

2.5.2 Ovine

No published reports on the endogenous or post-administration concentrations of boldenone or related compounds in the ovine were found. However, some Australian data from a national monitoring programme (origin of the animals not therefore controlled for steroid abuse) suggest that epiboldenone may be endogenous in the ovine. Of 961 ovine urine samples analysed for boldenone and epiboldenone, two (0.2%) had values above the LOD for epiboldenone with concentrations of 5.9 and 17 ng ml^{-1} respectively but no samples had values above the LOD for boldenone.[57]

2.5.3 Porcine

Only one published study relating to the occurrence of boldenone in the porcine was found. In a study on boars, barrows, gilts and sows, Poelmans *et al.*[101] reported the concentrations of endogenous boldenone and 19-nor-androgens in muscle, liver, kidney, testicles and urine using GC-MS and LC-MS methods. Calibration ranges and LODs were not reported, but the resulting concentration ranges in the different tissues were reported. The results from this study are presented in Table 2.5 along with the nandrolone data discussed above. In addition to the data in Table 2.5, the authors looked at concentrations of boldenone in an intersex animal, with muscle, liver, kidney, urine and testicles found to contain no boldenone above the LOD. More studies to determine the endogenous occurrence of boldenone in the porcine are clearly warranted.

2.6 Endogenous Occurrence of Oestradiol-related Compounds

Oestrogenic compounds are generally those which, among other notable effects, stimulate mitotic activity in tissues of the female genital tract. The most important naturally occurring oestrogens are 17β-oestradiol and oestrone, while 17α-oestradiol (epioestradiol) and oestriol are generally considered as metabolites with lower activity[102] (with the latter predominantly formed and released during pregnancy[3]). Normally, the ovaries are the major source of oestrogen production and 17β-oestradiol the major product.[25] However, during pregnancy the placenta becomes the major source.[25] Production rates are high in cows during late pregnancy (several hundred mg every 24 hours).[103] The male gonad can also produce appreciable amounts of oestrogens, especially in the boar and the stallion.[25] Ruminants metabolise 17β-oestradiol to the 17α-isomer, which is the main urinary metabolite and possesses very little oestrogenic activity.[103] The majority of 17β-oestradiol is excreted in the bile.[103]

2.6.1 Bovine

There have been several studies reporting on the concentrations of oestradiol and its metabolites in bovine. Every matrix has been studied to some extent although for some there are insufficient data to draw any useful conclusions.

One of the most comprehensive individual studies looking at oestrogen concentrations in the bovine was reported by Henricks and Torrance.[104] In the heifer, the lowest concentrations of 17β-oestradiol were observed in muscle and three-fold higher concentrations were seen in the liver and kidney while in steers there was no difference in the concentrations in the three different matrices. It was also noted that concentrations of oestrogen in muscle, kidney and liver did not vary with oestrous in the heifer, unlike the uterine endometrium where concentrations varied considerably depending on the stage of oestrous.[104]

Arts *et al.* performed carefully controlled experiments to establish reference values of oestradiols in plasma, faeces and urine of male and female veal calves, which indicated that plasma and urine are the preferred matrices for analysis.[45] A subsequent study by the same authors[51] reported the effect of age on male and female plasma and urinary oestrogen levels. Median plasma oestradiol levels in male calves at 15, 22 and 28 weeks were reported as below 20 pg ml^{-1} and in female calves the same levels were seen at 15, 24–25 and 28 weeks. In urine, the range of urinary oestradiol in male calves was ND to 100 pg ml^{-1} at 15 weeks and ND to 200 pg ml^{-1} at 28 weeks, while for epioestradiol the median concentration was 1200 pg ml^{-1} with a range of ND to 7100 pg ml^{-1} at

15 weeks. Levels had fallen to 200 pg ml^{-1} with a range of ND to 5700 pg ml^{-1} at 28 weeks. In female calves, urinary oestradiol concentrations were ND to 60 pg ml^{-1} at 15 weeks and ND to 100 pg ml^{-1} at 28 weeks, while for epioestradiol the median was 800 pg ml^{-1} with a range of ND to 4200 pg ml^{-1} at 15 weeks and 1500 pg ml^{-1} with a range of ND to 5400 pg ml^{-1} at 28 weeks. The authors commented that female calves reach first oestrus, which is accompanied by increased oestrogen production, at about 30 weeks of age, which may explain the relatively low concentrations described above. These results indicate the wide range of concentrations found in animals and illustrates the problems in setting threshold values.

A rigorous quantitative metabolism study regarding the occurrence of oestrogens in the bovine showed that by comparing the absolute concentrations of the different analytes in different matrices from "blank" animals, the concentrations of both oestradiol and epioestradiol in all sexes were highest in bile, followed by urine, and then serum.[105]

Table 2.6 summarises oestradiol concentrations derived from the Joint FAO/WHO Expert Committee on Food Additives (JECFA) review on the concentrations of testosterone, progesterone and oestradiol in edible cattle tissues.[103]

Effects with season have been studied by Sartori *et al.*,[106] who found that mean peak serum oestradiol after PGF2a induced oestrus was higher in heifers ($P < 0.01$) than in lactating cows during the summer but not during the winter. Unfortunately, no studies relating to variations in oestrogen levels with time of day were found.

Diseases of the hepatic tissue are well known to affect the concentrations of oestrogens and other steroids in man. Lopez-Diaz *et al.*[107] have shown that liver disease can also affect oestrogen concentrations in the bovine. The authors looked at the effects of infection with infected flukes of the *Fasciloa hepatica* species on the concentrations of serum oestradiol over a prolonged period of time in heifers who were four months old and pre-pubertal at the start of the experiment. Mean basal oestradiol concentrations in control heifers before first oestrus were 4.4 pg ml^{-1} while in infected heifers they were 10 pg ml^{-1}. The mean concentrations disguise a highly variable trend over time and in one infected heifer oestradiol peaked at 380 pg ml^{-1}. Oestradiol was also apparently higher after first oestrus, although no data were given to support this assertion. Increased oestradiol concentrations were proposed to be related to reduced hepatic clearance of oestradiol in the infected animals. If such liver infections are commonplace among herds, then this could have a significant negative impact on the validity of a plasma oestradiol threshold, However, if reduced hepatic clearance is the causative mechanism responsible for the increased plasma oestradiol concentrations, then one might expect that urinary concentrations might be lower than in non-infected animals.

It has been shown that the bovine diarrhoea virus does not affect peak plasma oestradiol concentrations in females after artificial synchronisation of oestrus, but does prolong the duration of high concentrations such that three days after synchronisation device removal, concentrations of oestradiol were 2.2 pg ml^{-1} in infected animals relative to 0.72 pg ml^{-1} in controls.[108] It has also

Table 2.6 Summary of oestradiol concentrations (in pg^{-1}) in cattle tissues summarising data from Arnold.[103]

	Muscle			*Fat*		
	min	*Range of medians*	*max*	*min*	*Range of medians*	*max*
Calves (steers)	0.36	0.47–2.18	8.79	1.68	2.6–4.59	22.9
Calves (heifers)	0.8	1.16–8.1	17.3	1.58	2.39–5.25	24.7
Heifers, non-pregnant	1.05	3.41	35.2	3.93	9.19	71.5
Heifers, pregnant	4.3	11.3–12.8	31.6	11.5	31.8–38.8	77.5
Steers	0.34	0.12–44.3	61	0.10	1.2–154	224
	Kidney			*Liver*		
	min	*Range of medians*	*max*	*min*	*Range of medians*	*max*
Calves (steers)	0.28	0.91–2.9	16.8	1.05	1.43–3.71	4.38
Calves (heifers)	0.99	1.22–6.85	8.54	1.16	1.48–7.51	8.22
Heifers, non-pregnant	0.5	1.97	11	0.48	1.00	3.89
Heifers, pregnant	63.8	99.6–120	258	12.1	49.9–53.1	26.7
Steers	0.55	1.53	2.78	0.36	0.87	1.72

been reported that 25% of cows suffer ovarian cysts and that analysis of such cows (calving at least six weeks prior to analysis) showed that they had plasma oestradiol levels that fell within a previously defined "normal" range.[47] Moura *et al.*[50] reported that oestradiol concentrations in maturing bulls were only slightly lower in those animals suffering spermatic arrest compared with controls and that this effect only reached statistical significance ($P < 0.05$) at two and three months of age. The effect of age on female plasma oestrogen concentrations was studied by Nakada *et al.*,[53] who found that mean oestradiol at birth was 23 $pg\,ml^{-1}$, dropping to a minimum of 0.52 $pg\,ml^{-1}$ after a week and then ranging from 1.0 $pg\,ml^{-1}$ to 2.0 $pg\,ml^{-1}$ until 20 weeks of age, after which concentrations increased to a maximum of 2.1 $pg\,ml^{-1}$ at puberty (mean age to puberty was 43.3 weeks, with a range of 38–55 weeks). However, it was shown by Gabai *et al.*[109] that taking plasma samples from different sites in the body showed a variation. They reported that mean oestradiol in jugular vein plasma of non-pregnant versus 250–280 day pregnant heifers were 7.34 and 88 $pg\,ml^{-1}$ respectively, while corresponding values for subcutaneous abdominal vein plasma were 9.9 and 290 $pg\,ml^{-1}$ respectively.

An approximate rank order for oestradiol concentrations in various matrices is hair > faeces > urine > liver ~ kidney > fat > plasma > muscle. An approximate rank order for levels of epioestradiol is faeces > urine > liver > kidney > plasma > fat > muscle > hair and for oestrone is liver ~ muscle ~ fat ~ kidney > plasma. There were no data for oestrone in urine, hair or faeces. In terms of absolute values, epioestradiol was the predominant metabolite in urine and faeces, while oestradiol was ten-fold lower in both. In plasma, oestradiol and epioestradiol were present at similar concentrations, which were around three-fold higher than those of oestrone. In fat, concentrations of oestradiol, epioestradiol and oestrone were similar. In liver, concentrations of oestradiol, epioestradiol and oestrone were similar. In kidney, concentrations of oestradiol and epioestradiol were similar and were both around two-fold higher than oestrone. In muscle, concentrations of oestrone were around two-fold higher than oestradiol and around ten-fold higher than epioestradiol. In hair, concentrations of oestradiol were highest, while epioestradiol was not detected. There were no data for oestrone in hair. However, it must be stressed that some positions within these rankings may be caused by biases in the amount of information reported for each steroid in different matrices.

Milk products are rich sources of steroids.[3] The hormone pattern resembles that of meat from female cattle. The main oestrogen in milk is epioestradiol followed by oestrone and 17β-oestradiol (ref. 68 in [3]). Processing, such as heating or churning, has no effect on hormone patterns, although cheese ripening does.

Bulls and heifers were found to have similar ranges of mean fat oestradiol, and these were around two-fold higher than in steers. Mean concentrations of oestrone were similar between steers, heifers and bulls, but there were insufficient data for a comparison of epioestradiol. In hair, mean oestradiol concentrations were slightly higher in heifers than in bulls, and were around 20-fold higher than those of steers. Ranges of mean muscle oestradiol were similar among heifers, bulls and steers, while oestrone was similar among heifers and bulls, but around two-fold lower in steers. Mean liver oestradiol concentrations were similar among heifers and bulls, but at least two-fold lower in steers. Oestrone was similar among heifers, bulls and steers while the only data for epioestradiol in steers and heifers suggests concentrations in heifers are around 30-fold greater than steers. In the kidney, mean oestradiol concentrations were highest in heifers, around two-fold lower in bulls and around four-fold lower in steers. Oestrone was similar in bulls and steers, but both were at least three-fold lower than that in heifers. Epioestradiol concentrations in heifers were around four-fold higher than those of steers. There were insufficient data to compare oestrogen concentrations by sex in faeces or bile. In the majority of cases there were either insufficient or no data to allow epiestradiol comparisons. Ranges of mean urinary oestradiol and epioestradiol concentrations were similar between heifers and bulls. No urinary oestrogen data were available for steers or heifers over different stages of the oestrous cycle or for oestrone.

2.6.2 Ovine

Endogenous concentrations of oestrogen in the ewe are somewhat governed by oestrous, which occurs at different times of the year depending on the degrees of latitude of residence of the sheep and whether the animals are wild, feral or domesticated.[110] Yuthasastrakosol *et al.* observed that there were random fluctuations of oestrogen concentrations deviating from a mean concentration of $4.40 \pm 0.1\,\mathrm{pg\,ml^{-1}}$ during anoestrus and that the mean concentration between first and second oestrous was $5.2 \pm 0.3\,\mathrm{pg\,ml^{-1}}$.[111]

2.6.3 Porcine

A study by Claus *et al.* showed that mean endogenous concentrations of 17β-oestradiol in pigs was $0.087\,\mathrm{ng\,ml^{-1}}$.[112] Claus *et al.* also measured the concentrations of conjugated and unconjugated oestrogens in peripheral blood plasma and seminal plasma in the boar, finding higher concentrations of the conjugated compared to the unconjugated concentrations.[113] They also consider seasonal effects, with higher concentrations observed in the fall, hence the large variations in the results shown. For blood plasma, these were $0.24 \pm 0.20\,\mathrm{ng\,ml^{-1}}$ and $17.01 \pm 14.94\,\mathrm{ng\,ml^{-1}}$ for the unconjugated and conjugated respectively, and in the seminal plasma $0.73 \pm 0.78\,\mathrm{ng\,ml^{-1}}$ and $5.89 \pm 4.95\,\mathrm{ng\,ml^{-1}}$ for the unconjugated and conjugated oestrogens respectively.[113] They also note that the concentrations of unconjugated oestrogens in whole semen are a factor of two higher than in seminal plasma.[113] In a second study on boar semen, the mean concentrations were found to be $55.4 \pm 8.4\,\mathrm{ng\,ml^{-1}}$ for unconjugated total oestrogens and $35.8 \pm 6.0\,\mathrm{ng\,ml^{-1}}$ for conjugated total oestrogens. The authors noted that of the unconjugated oestrogens 77% were 17β-oestradiol and 23% were oestrone.[114] It was also noted that the concentrations of unconjugated oestrogens in the tubular fluid are 25% higher than in the testicular fluid.[114]

2.7 Endogenous Occurrence of Progesterone-related Compounds

Progesterone is the predominant mammalian progestagen (a substance which favours gestation), the function of which is to prepare and maintain the lining of the female reproductive tract for pregnancy. It can reduce the oestrogenic effects of simultaneously administered oestrogens,[3] as well as reducing testosterone-induced aggressive behaviour by inhibiting the formation of testosterone metabolites.[115] Progesterone is an intermediate in the production of other steroid hormones and the adrenals can release significant amounts if stimulated.[25] It is an important precursor for the biosynthesis of androgens and corticosteroids and is also present in males and castrates. Conversion to androgens and excretion *via* faeces is an important route in ruminants.[103] In the cow, goat and sow, the corpus luteum is the main source during the oestrous cycle and pregnancy,[103] but in other species such as the sheep, the placenta becomes the prime site for progesterone synthesis in the latter stages of pregnancy.[25]

2.7.1 Bovine

There was little information available to compare the concentrations of progestagens between sexes, but the data found suggested that concentrations were higher in heifers compared with bulls and steers, but it is difficult to say by exactly how much due to the dearth of data in the male animals. Plasma/serum progesterone concentrations in heifers varied up to ten-fold with the stage of the oestrous cycle, while concentrations in other tissues varied by somewhat lesser extents. There were insufficient data to compare pregnenolone and 17α-hydroxyprogesterone between sexes in most cases, but concentrations of each in the muscle of bulls and steers were found to be similar.[57]

Progesterone is usually found in female cattle, although it can exist in the male, with values as high as 43.7 ng g^{-1} in tissue and can be stored in fat, with concentrations of ~17 ng g^{-1} observed in the fat of heifers and concentrations as high as 100–400 ng g^{-1} in pregnant cows.[116] The degradation products pregnene-20α-ol-3-one and 17α-hydroxyprogesterone have also been observed for those samples which had high concentrations of progesterone, predominantly from female cows[116] or directly reduced at C20 and/or C3, together with saturation of ring A, resulting in many possible isomers of pregnanediol.[25] The major metabolites of progesterone include allopregnanolone, 5α-pregnane-3,20-dione and fatty acid esters of allopregnanolone.[117]

The effect of age on male and female plasma and urinary progesterone concentrations was studied by Arts *et al.*, whose findings are shown in Table 2.7.[51] The authors estimate that first oestrus is reached at about 30 weeks

Table 2.7 Summary of reference values of naturally occurring steroids in urine from Arts. *et al.*[50]

	Reference value ($\mu g\,l^{-1}$) of naturally occurring total steroids in urine							
	Male Calves				*Female Calves*			
	15 weeks		*28 weeks*		*15 weeks*		*28 weeks*	
	Median	*Range*	*Median*	*Range*	*Median*	*Range*	*Median*	*Range*
17β oestradiol	<0.05	<0.05–0.1	<0.05	<0.05–0.1	<0.05	–	<0.05	–
17α oestradiol	1.2	0.3–5.0	2.0	0.6–5.0	0.8	<0.3–2.5	1.5	<0.4–3.8
Progesterone	1.0	<0.6–2.0	2.2	0.8–5.5	0.9	<0.6–2.5	3.2	1.4–7.0

of age, but concentrations were not measured past this age. The effect of age on female Holstein-Friesian plasma progesterone concentrations has also been studied[53] and shows that mean progesterone from birth to puberty ranged from 0.05 to 0.18 ng ml^{-1}, rising to 3.0 ng ml^{-1} one week after puberty (onset of puberty ranged from 38 to 55 weeks, with the mean age at 43.3 weeks). The mean age to first oestrous in Korean Native heifers was 344±15 days (49.1 weeks)[118] where plasma progesterone concentrations before puberty were around 0.5 ng ml^{-1}, varying after first oestrous between 0.5 and 7.5 ng ml^{-1} depending upon the stage of the cycle.

There were insufficient data to compare the effect of age on progestagens in faeces, liver, kidney, bile, muscle, hair or fat.

Effects with season have been studied by Sartori *et al.*[106] Mean serum progesterone concentrations seven days after PGF2a induced oestrous were higher ($P < 0.01$) in heifers than in lactating cows during the summer but not during the winter.

Dobson *et al.*[47] reported that 25% of cows suffer ovarian cysts and that analysis of such cows (calving at least six weeks prior to analysis) showed that their plasma progesterone concentrations fell within a previously defined "normal" range. No reports on the effect of other factors known to increase the progestagen output in other species were found *i.e.* congenital adrenal hyperplasia.

An effect of diet on steroid concentrations is not entirely unexpected, as nutrient intakes are intimately tied in with reproductive function. This is exemplified by the observation that leptin concentrations, which are correlated with reproductive function, increase with calorie intake in many species.[119] Whether or not these differences merely demonstrate a permissive action of a healthy diet on normal steroid concentrations, or whether they actually signify a trend that would be continued if doses of the substances involved were increased remains to be established. Other diet studies have included the effect of increased selenium[120] and prilled long-chain fatty acids concentrations[121] on plasma progesterone concentrations. Selenium appeared to increase progesterone concentrations in plasma with mean peak concentrations recorded as 6.3 ng ml^{-1} in the selenium supplemented group compared with 5.3 ng ml^{-1} in the controls.[120] When 5% prilled long-chain fatty acids were incorporated into the diet, mean plasma progesterone concentrations were higher ($P < 0.05$)

in the supplemented group during the mid/late luteal phases (concentrations of 7.7 and 7.6 ng ml^{-1} respectively) of the second and third oestrous cycles compared with the controls (concentrations of 6.5 and 6.6 ng ml^{-1} respectively).

An approximate rank order of absolute concentrations for progesterone (excluding milk) is fat ~faeces > urine > muscle > plasma/serum > liver ~kidney. Data for progesterone in bile or hair were not provided. The differences between the different milk matrices were so large that any differences between pregnant and non-pregnant animals would seem to be insignificant in comparison. An approximate rank order for progesterone in milk matrices as milk fat ≫ whole milk > skimmed milk > milk whey.[57] The concentrations in milk fat were around ten-fold higher than fat or faeces (the matrices with the second highest concentrations), while concentrations in whole milk were similar to those in fat or faeces.

There was insufficient data to compare 17α-hydroxyprogesterone concentrations between the majority of matrices, but reported concentrations were similar in plasma/serum and muscle. Absolute concentrations of pregnenolone were much higher than those of progesterone in bull plasma, but concentrations of the two steroids in steer muscle were similar. These results are in contrast to another study which found that the muscle of steers contained higher concentrations of pregnenolone ($P < 0.05$) but not progesterone or 17α-hydroxyprogesterone compared with bulls.[122] There were no data for the two steroids in the heifer. Absolute concentrations of 17α-hydroxyprogesterone were always lower than those of progesterone or pregnenolone in bull and steer matrices. Again, it must be stressed that some positions within these rankings may be caused by biases in the amount of information reported for each steroid in different matrices.

Pregnenolone was found to be present in all samples analysed in a study by Hartwig,[116] who reported concentrations of 0.5 to 6.5 ng g^{-1} in meat, with females showing the highest values. However, when other precursors, such as androstenedione were considered, there did not appear to be any gender preference.[116] However, when the progesterone/pregnenolone ratio was investigated, it was found that in bulls and in most of the steers the ratio was <0.5 but in heifers all ratios were greater than three, with the majority greater than five.[116] This finding may be used to identify the sexual origin of meat.

Data for the conjugation status of progestagens and pregnenolone in different tissues were lacking.

2.7.2 Ovine

There was very little information on the concentrations of progesterone in sheep, but it appears that concentrations remain low during the non-breeding season.[110] Yuthasastrakosol *et al.* monitored the hormone concentrations between first and second oestrus and found that progesterone concentrations were 0.11±0.01 ng ml^{-1} 25 days before anoestrus, rising on day 12 of anoestrus.[111]

2.7.3 Porcine

There are very few data available on the progesterone concentrations in pigs. However, Bell *et al.*[123] studied the concentrations during the oestrous cycle. They found that in sows, progesterone concentrations in the ovarian vein were between 0.8 and 24.9 $ng\,ml^{-1}$ on day 2, increasing to between 125.8 and 2978.8 $ng\,ml^{-1}$ on day 16 before falling slightly to between 19.4 and 2030.8 $ng\,ml^{-1}$ on day 19. For gilts, progesterone concentrations in the posterior vena cava were between 0.2 and 82.8 $ng\,ml^{-1}$ on day 2, increasing to between 10.7 and 1100.0 $ng\,ml^{-1}$ on days eight to ten and falling to between 0.6 and 113.0 $ng\,ml^{-1}$ on days 16/17.[123]

2.8 Endogenous Occurrence of Cortisol-related Compounds and Their Effect on Other Endogenous Hormones

Cortisol is the major endogenous glucocorticoid that could be subject to abuse in food-production. It is ubiquitous in all mammalian species and is subject to both diurnal variation (high in the morning and low at night) and increases with both stressful situations and food deprivation.

Cortisol and cortisone concentrations in beef cattle have been studied by several groups in Europe. Fritsche *et al.* determined concentrations in different muscles from bulls and found median concentrations of cortisone to be 1.17, 1.87 and 1.47 $ng\,g^{-1}$ in *longissimus dorsi* (LD) *semitendinosus* (ST) and *extensor carpi ulnaris* (ECU) muscles respectively,[124] while in steers these concentrations were 1.65, 1.44 and 2.33 $ng\,g^{-1}$ respectively. For cortisol, the median concentrations recorded in bulls were 4.37, 3.82 and 4.87 $ng\,g^{-1}$ respectively and in steers 5.57, 4.76 and 8.58 $ng\,g^{-1}$ respectively. The authors suggest that, at least in muscle, equilibrium hormone concentrations are reached such that the rates of influx and metabolism are balanced. Cortisol and cortisone are more polar than other steroids and therefore do not accumulate in fat.[124]

A study by Hartmann and Steinhart found that the variability between replicates, as demonstrated by the % CV, of cortisone reached almost 40% at a concentration of 0.6 $ng\,g^{-1}$. They suggest that this compound is non-homogeneously distributed through tissue, making accurate analysis very difficult.[125]

The effect of cortisol concentrations on the concentrations of other steroids has also been an area of some study. For example, Ryan *et al.*, who studied the effect of cortisol in steroid production in prepubertal gilts found that progesterone and androstenedione synthesis was depressed in the presence of cortisol, but that oestradiol and androsterone synthesis was increased in follicles.[126]

2.9 Variation with Pregnancy and Oestrous

In discussing the concentrations of hormones in meat-producing animals, the effects of oestrous and pregnancy cannot be ignored. The discussion of steroid concentrations during pregnancy and oestrus will focus on the bovine as this is the species where the majority of information is available but where information is available for other species, this will be referred to. There has also been some discussion of the effects of pregnancy in other sections of this review. Oestradiol concentrations have been shown to vary considerably depending on the stage of oestrous or pregnancy in heifers and the tissue being studied. Plasma/serum oestradiol concentrations in heifers varied up to ten-fold with the stage of the oestrous cycle, while concentrations in other tissues varied by somewhat lesser extents. A relatively large number of studies reported oestrogen concentrations during pregnancy, all of which show a progressive increase with term and with the number of foetuses being carried. In a study by Dobson *et al.*,[47] mean oestrous cycle maximum plasma oestrone, oestradiol and epioestradiol concentrations in non-pregnant cows were 6, 13 and 30 pg ml^{-1} respectively. However, concentrations in pregnant cows 14 days before parturition were markedly higher with values of 880, 400 and 460 pg ml^{-1} respectively for the three steroids. These concentrations then rose further so that the day before parturition values were 3000, 1100 and 850 pg ml^{-1} respectively.

Hirako *et al.*[127] commented that while bovine plasma concentrations of oestrone sulfate are lower than many other species during the normal female oestrous cycle, concentrations increased significantly during pregnancy. The authors also showed that concentrations of oestrone sulfate were increased in animals carrying two as opposed to one foetus. Mean oestrone sulfate concentrations ranged from 1 to 6 pg ml^{-1} in animals carrying one and two foetuses respectively until day 50, rising to 45 and 79 pg ml^{-1} at day 100. Similarly, mean oestrone ranged between 2 and 4 pg ml^{-1} until day 80 in all animals, rising to 9 and 19 pg ml^{-1} at day 100 in animals carrying one and two foetuses respectively. Mean oestradiol ranged between 1.5 and 4 pg ml^{-1} until day 80 in all animals, rising to 4 and 7 pg ml^{-1} in animals carrying one and two foetuses respectively.[128] They also showed that there was not a significant difference between Holstein and Japanese Black breeds. The authors defined "day zero" as the first day of standing oestrous. In addition to the progressive increase in concentrations during pregnancy described above, Janowski *et al.*[129] showed that plasma epioestradiol reached a peak of 1150 pg ml^{-1} at 6 days prior to parturition, while oestradiol, oestrone and epioestradiol peaked around the time of parturition with values of 200, 970 and 31,000 pg ml^{-1} respectively.

The variation in progestagens, glucocorticoids, oestradiol and oestrone concentrations in blood serum from heifers from 26 days pre-partum to nine days post-partum was investigated.[130] Results are summarised in Table 2.8. It can be seen that progesterone concentrations fall rapidly just before parturition and remain at 0.6 ng ml^{-1} after birthing.

Oestradiol and oestrone concentrations rise in the week before parturition and then fall at birth, declining further in the days following. Glucocorticoids, however, increase in concentration before parturition, reaching a maximum at that time. They then return to pre-partum concentrations after birth. The authors cite a number of other references to similar observations, with hormone

Table 2.8 Summary of mean blood serum hormone concentrations during late pregnancy, parturition and early lactation. Concentration levels are $ng\,ml^{-1}$ Taken from Smith *et al.*[129]

	Pre-parturition					*Parturition*	*Post-parturition*		
	−26 days	*−5 days*	*−2 days*	*−1 day*	*−0.5 days*		*0.5 days*	*1 day*	*9 days*
Progesterone	10.1	–	7.6	3.0	0.6	–	–	–	0.6
Glucocorticoids	5.0	–	–	6.4	10.3	16.7	5.1	–	5.1
Oestradiol	0.032	0.15	0.293	–	–	0.18	–	0.052	0.028
Oestrone	0.218	–	2.256	–	–	0.726	–	0.115	0.014

concentrations beginning to drop rapidly 24–48 hours before parturition. They describe the relationship between the various hormones studied and parturition thus: "The first major change in serum steroid hormones is a ten fold increase in estrogens during the month before parturition, to a peak at about two days before parturition. Blood estrogens reach basal postpartum values within three days after parturition. Progesterone remains high until three days before parturition and falls principally during the last two days of pregnancy to values at parturition typical of estrus. The parturient rise in glucocorticoids occurs shortly after the fall in progesterone. Glucocorticoids peak at parturition and return to basal values within 12 hours. The large increase in estrogen probably increases contractility of the uterus during the final weeks of gestation, and the precipitous decrease in progestins during the 48 hour prepartum leaves the uterus under estrogen dominance at a time when coordinated uterine contractions begin in cattle."[130]

When urine was analysed, much the same trends were seen as for hormones in plasma during late pregnancy.[131] Total oestrogen (oestradiol plus epioestradiol) concentrations increase until parturition and then rapidly decrease. The authors noted that epioestradiol concentrations increase to 76% of the total oestrogen content at parturition and to 93% 0.5 day after calving, compared to oestrone which is 21% and 5% respectively.[131] The authors also noted that the length of gestation also appeared to have an effect on hormone concentrations pre-partum, although this could also be an effect of the different breeds and was not thought to be statistically significant.

Existing EEC guidelines for positive oestradiol decision limits in the bovine do not rely on sex or age status, but do rely on gestation status. These guidelines currently state that the decision limit in plasma for non-pregnant females and male animals of any age is $0.04\,ng\,ml^{-1}$.[132]

Hamudikuwanda *et al.*[133] studied the concentrations of progesterone in plasma and subcutaneous tail fat in pregnant and subsequently lactating Holstein heifers. Two samples were taken at different times during late pregnancy and then three samples were taken at different times after parturition, during lactation. Mean progesterone concentrations in the first samples taken during late pregnancy were $230\,ng\,g^{-1}$ in fat and $7.1\,ng\,ml^{-1}$ in plasma. Concentrations in the second samples taken during late pregnancy were $200\,ng\,g^{-1}$

in fat and 5.6 ng ml^{-1} in plasma. Mean plasma progesterone concentrations in the first, second and third samples taken during subsequent lactation were 0.1, 1.7 and 3.5 ng ml^{-1}, while corresponding values in fat were 14, 11 and 74 ng g^{-1} respectively. There was a significant decrease in concentrations in both tissues following parturition, but concentrations then rose back to around half those during pregnancy in the later samples (possibly due to a resumption to oestrous cycles). The concentrations of progesterone in the plasma of pregnant and subsequently lactating heifers, this time in crosses of the Holstein-Friesian x Red Sindhi breeds have been studied by Chaiyabutr *et al.*,[134] who showed that progesterone concentrations were high 28 days pre-partum, dropped after parturition and then increased to concentrations slightly less than those seen in pregnancy by 210 days after parturition.

A study of the difference in plasma progesterone levels between a pregnant and a cycling Japanese Black heifer showed that in the pregnant cow, progesterone peaked at 4.7 ng ml^{-1} at 175 hours prior to parturition and then dropped gradually to less than 1 ng ml^{-1} around the time of parturition. In the cycling cow, progesterone during the oestrus cycle range from a low of less than 1 ng ml^{-1} to a high of 8 ng ml^{-1}.[135] Isobe *et al.*[136] carried out a study to assess whether faecal progesterone concentrations could be an aid to diagnosing pregnancy in crosses of the Holstein-Friesian x Japanese Black crossed breeds. Mean progesterone in non-pregnant animals peaked at 65 ng g^{-1} whereas pregnant animals peaked at 110 ng g^{-1}. From day 19–24 after oestrus in the non-pregnant animals or 19–24 days after artificial insemination in the pregnant animals, progesterone was greater ($P < 0.05$) in pregnant animals compared with non-pregnant heifers.

Plasma progesterone concentrations between repeat breeder heifers were compared to those in virgin heifers over an average of 5.3 oestrous cycles.[137] There was a trend toward lower progesterone at day zero of the cycle and higher progesterone at the peak of the cycle in virgin heifers, but this did not reach statistical significance ($P < 0.05$).

In a 2003 paper reviewing studies that have linked a genetic selection for milk yield with a reduction in fertility of animals, Veerkamp *et al.*[138] cited studies by Eley *et al.*[139,140] that demonstrated increased plasma progesterone, but reduced plasma oestrone, in animals that were genetically selected for increased milk production. However, the authors comment that their own studies[141] and those by Crooker and Lucy[142] suggest that post-partum plasma oestradiol concentrations were no different between selected and un-selected animals, although progesterone concentrations were reduced in selected animals.

No data were available on the concentrations of testosterone, boldenone or their metabolites in any matrix from pregnant females, but they would be expected to be elevated relative to non-pregnant females in line with other steroids. However, mean plasma concentrations of DHEA and androst-5-ene-3β,17β-diol were found to be approximately three-fold higher in pregnant females.[109] Concentrations of nandrolone-related compounds during pregnancy in the bovine were covered in Section 2.4.

A study that noted increases in the uterine luminal corticosteroids cortisol and aldosterone during the oestrous cycle and early pregnancy in pigs was presented by Klemcke *et al.*[143] The authors note that similar changes in concentrations are seen for both cycling and pregnant gilts and suggest that the maternal adrenal cortex is the source of these compounds.

It has been reported that the newborn of the ovine, caprine and bovine display very high rates of 20α-hydroxysteroid dehydrogenase activity (acting on progestagens and corticosteroids) and that this activity diminishes rapidly (although not completely) with age, possibly due to the replacement of foetal with adult erythrocytes.[144]

2.10 Discussion and Conclusion

The mammalian body is a very complex system with many different compounds such as steroid hormones involved in maintaining homeostasis. The concentrations of these compounds depend on many factors including diet, sex, stage of oestrous, castration, age, disease and environment, and the balance between these compounds is very finely tuned. If anything should happen to upset this delicate balance then the rest of the body will try to compensate.

When endogenous hormones are considered, there is a plethora of precursors and metabolites, making the detection and quantitation of these compounds very complicated. The catabolic and anabolic hormones are present in both the male and female animals, but their concentrations can be significantly different depending on the many factors detailed in this review.

As well as providing an overview of the data currently available for various combinations of steroid/matrix/sex/age/gestate/castrate, this report highlights several analytical and physiological trends within the data. One of the most challenging aspects when analysing the data is the fact that many studies do not report all experimental details or results. Biases in sampling designs complicate meta-analysis of the literature, such that there are large numbers of studies relating to longitudinal oestrus and pregnancy cycles, and fewer studies relating to control of abuse. Analytically, the use of different detection techniques leads to variation in reported steroid concentrations. The differential use of sample preparation steps such as steroid hydrolysis and the addition of matrix stabilisers further adds to the artifactual variation within the data. Physiologically, it would seem necessary to separate animals based upon, at the very least, their age, sex, castration, gestation, disease and medication status otherwise the variation under these different conditions may lead to irrelevantly high thresholds. However, there would appear to be balance between a physiological need to separate animals and a statistical requirement for a large enough population on which to base threshold calculations.

Several questions have been raised, including some unusually high steroid concentrations under specific physiological conditions and a relevant dearth of information regarding some steroids in different species (*i.e.* boldenone/ nandrolone). Furthermore, continuously improving sensitivities of analytical detection, as well as artificial selection within certain animal populations, give rise to a continual evolution of steroid concentrations and suggest constant re-evaluation of detection approaches is necessary. An ideal situation might therefore be analogous to human sports: the use of a flexible threshold value/ratio as a screening strategy and the use of more laborious but definitive techniques such as carbon isotope ratio mass spectrometry for confirmation.

At the time of this book going to press, additional work is underway by the authors to obtain more information on the natural range of steroid hormone concentrations in matrices from meat-producing animals. The goal is to use these data to guide screening and/or confirmatory thresholds for controlling the abuse of natural steroids in mammalian food production.

However, it appears that the more research is performed in this area, the more questions are raised and the more complex the situation appears. It may be that common thresholds satisfactory to every meat-producing country will never be realised.

Acknowledgements

This article first appeared in a shortened form in the following article and is reproduced with permission from Taylor and Francis Ltd, (http://www.tandf.co.uk/journals): Scarth et al. 2009. Presence and metabolism of endogenous androgenic–anabolic steroid hormones in meat-producing animals: a review. Food Additives and Contaminants: Part A. vol 26(5). pg. 640–671.

References

1. Oxford English Dictionary online, Oxford University Press, 2006, http://www.askoxford.com/concise_oed/sexhormone?view=uk, accessed 26 March 2009.
2. Merriam-Webster online dictionary, http://www.m-w.com/dictionary/sex + hormone, accessed 26 March 2009.
3. S. Fritsche and H. Steinhart, *Eur. Food Res. Tech.*, 1999, **209**, 153.
4. European Commission, http://eur-lex.europa.eu/LexUriServ/LexUriServ.do?uri=CELEX:31996L0022:EN:HTML, 1996, accessed 26 March 2009.
5. European Commission, http://eur-lex.europa.eu/LexUriServ/LexUriServ.do?uri=CELEX:31996L0023:EN:HTML, 1996, accessed 26 March 2009.
6. G. Balizs and A. Hewitt, *Anal. Chim. Acta*, 2003, **492**, 105.
7. European Commission, http://eur-lex.europa.eu/LexUriServ/LexUriServ/site/en/oj/2002/l_22120020817en00080036.pdf, 2002, accessed 26 March 2009.
8. A. A. M. Stolker and U. A. T. Brinkman, *J. Chrom.*, 2005, **1067**, 15.
9. R. J. Heitzman, in *Veterinary Drug Residues: Residues in Food Producing Animals and Their Products: Reference Materials and Methods*, Blackwell Scientific, Brussels and Luxembourg, 2nd EC Report EUR 15127 ed., 1994.

10. The United States Mission to the European Union, http://useu.usmission.gov/Dossiers/Beef_Hormones/Mar31_08_WTO_Dispute_Panel.asp, 2008, accessed 26 March 2009.
11. W. Velle, in *Hormones in animal production, animal physiology, feed additives, FAO Animal Production and Health Paper (FAO)*, no. 31, ed. Joint FAO/WHO Expert Committee on Food Additives, 1982, pp. 1–24.
12. Food and Drug Administration, http://www.fda.gov/ohrms/dockets/98fr/2005d-0219-gd0001.pdf, 2005, accessed 26 March 2009.
13. L. D. Bowers and Sanaullah, *J. Chrom. B Biomed. Sci. Appl.*, 1996, **687**, 61.
14. M. W. F. Nielen, J. J. P. Lasaroms, P. P. J. Mulder, J. Van Hende, J. van Rhijn and M. J. Groot, *J. Chrom. B*, 2006, **830**, 126.
15. M. Gratacos-Cubarsi, M. Castellari, A. Valero and J. A. Garcia-Regueiro, *J. Chrom. B*, 2006, **834**, 14.
16. L. A. van Ginkel, R. W. Stephany, A. Spaan and S. S. Sterk, *Anal. Chim. Acta*, 1993, **275**, 75.
17. I. Hanukoglu, *J. Steroid Biochem. Mol. Biol.*, 1992, **43**, 779.
18. P. E. Joos and M. Van Ryckeghem, *Anal. Chem.*, 1999, **71**, 4701.
19. P. O. Edlund, L. Bowers and J. Henion, *J. Chrom. B Biomed. Sci. Appl.*, 1989, **487**, 341.
20. R. P. Martin, *Endocrinology*, 1966, **78**, 907–913.
21. W. Schanzer, *Clin. Chem.*, 1996, **42**, 1001.
22. H. F. De Brabander, S. Poelmans, R. Schilt, R. W. Stephany, B. Le Bizec, R. Draisci, S. S. Sterk, L. A. van Ginkel, D. Courtheyn, N. Van Hoof, A. Macri and K. De Wasch, *Food Addit. Contam.*, 2004, **21**, 515.
23. M. Metzler, *J. Chrom. B Biomed. Sci. Appl.*, 1989, **489**, 11.
24. H. Bi and R. Masse, *J. Steroid Biochem. Mol. Biol.*, 1992, **42**, 533.
25. W. Velle, in *Environmental Quality and Safety, Anabolic Agents in Animal Production*, ed. F. Coulston and F. Corte, Suppl. Vol. V, Thieme, Stuttgart, 1976, 159–170.
26. C. Aman, A. Pastor, G. Cighetti and M. de la Guardia, *Analytical and Bioanalytical Chemistry*, 2006, **386**, 1869.
27. E. Houghton, L. Grainger and M. C. Dumasia, *OMS*, 1992, **27**, 1061.
28. J. Scarth, Personal Communication, 2007.
29. H. L. J. Makin, D. B. Gower and S. Garai, in *Steroid Analysis,* Blackie Academic and Professional, 1st edn., 1995.
30. J. I. Mason, in *Genetics of Steroid Biosynthesis and Function,* Taylor and Francis, , 2002.
31. M. E. Hadley and J. Levine, in *Endocrinology,* Prentice Hall, 2006.
32. M. Groot, PhD thesis, University of Utrecht, The Netherlands, 1992.
33. R. Gaiani, F. Chiesa, M. Mattioli, G. Nannetti and G. Galeati, *Reproduction*, 1984, **70**, 55.
34. R. W. Stephany, S. S. Sterk and L. A. van Ginkel, Tissue levels and dietary intake of endogenous steroids and overview with emphasis on 17beta-estradiol, Euroresidue 5, 2004.
35. W. Challenger, Queen's University, Kingston, Ontario, Canada, 2008.

36. U. Wichmann, G. Wichmann and W. Krause, *Exp. Clin. Endocrinol.*, 1984, **83**, 283.
37. B. J. Johnson and B. J. Everitt, in *Essential Reproduction,* Blackwell Science, 2000.
38. A. E. Pusateri, M. E. Wilson and M. A. Diekman, *Biol. Reprod.*, 1996, **55**, 590.
39. E. S. Lopes Jr, J. F. Cruz, D. I. A. Teixeira, J. B. Lima Verde, N. R. O. Paula, D. Rondina and V. J. F. Freitas, *Vet. Res. Comm.*, 2004, **28**, 119.
40. L. A. Maclaren, *Reprod. Fertil. Dev.*, 1993, **5**, 261.
41. M. J. Rathbone, K. L. MacMillan, W. Jochle, M. P. Boland and E. K. Inskeep, *Crit. Rev. Ther. Drug Carrier Syst.*, 1998, **15**, 285.
42. R. Palme, P. Fischer, H. Schildorfer and M. N. Ismail, *Anim. Reprod. Sci.*, 1996, **43**, 43.
43. W. Korth to James Scarth, Personal Communication, 2008.
44. M. L. Scippo, P. Gaspar, G. Degand, F. Brose, G. Maghuin-Rogister, P. Delahaut and J. P. Willemart, *Anal. Chim. Acta*, 1993, **275**, 57.
45. C. J. M. Arts, M. J. van Baak and J. M. P. Den Hartog, *J. Chrom. Biomed. Appl.*, 1991, **564**, 429.
46. D. E. Gerrard, S. J. Jones, E. D. Aberle, R. P. Lemenager, M. A. Diekman and M. D. Judge, *J. Anim. Sci.*, 1987, **65**, 1236.
47. H. Dobson, J. E. Rankin and W. R. Ward, *Vet. Record*, 1977, **101**, 459.
48. E. T. Bagu, S. Cook, C. L. Gratton and N. C. Rawlings, *Reproduction*, 2006, **132**, 403.
49. R. Renaville, S. Massart, M. Sneyers, M. Falaki, N. Gengler, A. Burny and D. Portetelle, *Reproduction*, 1996, **106**, 79.
50. A. A. Moura and B. H. Erickson, *Theriogenology*, 2001, **55**, 1469.
51. C. J. M. Arts, M. J. van Baak, H. Van der Berg, R. Schilt, P. L. M. Berende and J. M. P. Den Hartog, *Archiv fur Lebensmittelhygine*, 1990, **41**, 58.
52. R. Angeletti, L. Contiero, G. Gallina and C. Montesissa, *Vet. Res. Comm.*, 2006, **30**, 127.
53. K. Nakada, M. Moriyoshi, T. Nakao, G. Watanabe and K. Taya, *Domest. Anim. Endocrinol.*, 2000, **18**, 57.
54. P. Plusquellec and M. F. Bouissou, *Appl. Anim. Behav. Sci.*, 2001, **72**, 1.
55. M. J. D'Occhio and W. J. Aspden, *Biol. Reprod.*, 1996, **54**, 45.
56. W. J. Aspden, N. van Reenen, T. R. Whyte, L. J. Maclellan, P. T. Scott, T. E. Trigg, J. Waleh and M. J. D'Occhio, *Domest. Anim. Endocrinol.*, 1997, **14**, 421.
57. J. Scarth, C. Akre, L. van Ginkel, B. Le Bizec, H. de Brabander, W. Korth, J. Points, P. Teale and J. Kay, Presence and metabolism of endogenous androgenic–anabolic steroid hormones in meat-producing animals: a review. *Food Addit. Contam.*, 2009, **26**, 640–71.
58. M. H. Fahmy, *Small Ruminant Research*, 1997, **26**, 267.
59. G. A. Langford, J. N. B. Shrestha, L. M. Sanford and G. J. Marcus, *Small Ruminant Research*, 1998, **29**, 225.
60. B. D. Schanbacher and J. J. Ford, *Biol. Reprod.*, 1979, **20**, 719.
61. C. Borque and I. Vazquez, *Small Ruminant Research*, 1999, **33**, 263.

62. H. J. D. Rosa, D. T. Juniper and M. J. Bryant, *Appl. Anim. Behav. Sci.*, 2000, **67**, 293.
63. T. J. Parkinson, K. C. Smith, S. E. Long, J. A. Douthwaite, G. E. Mann and P. G. Knight, *Reproduction*, 2001, **122**, 397.
64. F. Schwarzenberger, G. S. Toole, H. L. Christie and J. I. Raeside, *Acta Endocrinol.*, 1993, **128**, 173.
65. S. A. McCoard, T. H. Wise and J. J. Ford, *J. Endocrinol.*, 2003, **178**, 405.
66. C. S. Park and Y. J. Yi, *Anim. Reprod. Sci.*, 2002, **73**, 53.
67. S. Walker, O. W. Robison, C. S. Whisnant and J. P. Cassady, *J. Anim. Sci.*, 2004, **82**, 2259.
68. E. Houghton, J. Copsey, M. C. Dumasia, P. E. Haywood, M. E. Moss and P. Teale, *Biomed. Mass Spectrom.*, 1984, **11**, 96.
69. G. Maghuin-Rogister, A. Bosseloire, P. Gaspar, C. Dasnois and G. Pelzer, *Ann. Med. Vet.*, 1988, **132**, 437.
70. M. Vandenbroeck, G. Van Vyncht, P. Gaspar, C. Dasnois, P. Delahaut, G. Pelzer, J. De Graeve and G. Maghuin-Rogister, *J. Chrom. Biomed. Appl.*, 1991, **564**, 405.
71. A.-S. Clouet, B. Le Bizec, M.-P. Montrade, F. Monteau and F. Andre, *Analyst*, 1997, **122**, 471.
72. S. Sterk, H. Herbold, M. H. Blokland, H. J. Van Rossum, L. A. van Ginkel and R. W. Stephany, *Analyst*, 1998, **123**, 2633.
73. L. Dehennin, P. Silberzahn, A. Reiffsteck and I. Zwain, *Pathol. Biol.*, 1984, **32**, 828.
74. J. Van Hende, Postgraduate thesis, Ghent University, Belgium, 1995.
75. P. Van Eenoo, F. T. Delbeke, F. H. de Jong and P. De Backer, *J. Steroid Biochem. Mol. Biol.*, 2001, **78**, 351.
76. B. Le Bizec, I. Gaudin, F. Monteau, F. Andre, S. Impens, K. De Wasch and H. F. De Brabander, *Rapid Comm. Mass Spectrom.*, 2000, **14**, 1058.
77. B. de Geus, F. Delbeke, R. Meeusen, P. Van Eenoo, K. De Meirleir and B. Busschaert, *Int. J. Sports Med.*, 2004, **25**, 528.
78. J. Grosse, P. Anielski, P. Hemmersbach, H. Lund, R. K. Mueller, C. Rautenberg and D. Thieme, *Steroids*, 2005, **70**, 499.
79. E. Houghton, P. Teale and M. C. Dumasia, *Anal. Chim. Acta*, 2007, **586**, 196.
80. E. Daeseleire, A. De Guesquiere and C. Van Peteghem, *Anal. Chim. Acta*, 1993, **275**, 95.
81. H. F. De Brabander, J. Van Hende, P. Batjoens, L. Hendricks, J. Raus, F. Smets, G. Pottie, L. A. van Ginkel and R. W. Stephany, *Analyst*, 1994, **119**, 2581.
82. J. D. G. McEvoy, C. E. McVeigh, W. J. McCaughey and S. A. Hewitt, *Vet. Record*, 1998, **143**, 296.
83. J. D. G. McEvoy, W. J. McCaughey, J. Cooper, D. G. Kennedy and B. M. McCartan, *Vet. Q.*, 1999, **21**, 8.
84. G. Casson, M. Navaneethanan and J. Points, Is 17-alpha-19-nortestosterone endogenous in male sheep urine?, 5th International Symposium on Hormone and Veterinary Drug Residue Analysis, 2006.

85. S. Poelmans, K. De Wasch, H. Noppe, N. Van Hoof, S. Van Cruchten, B. Le Bizec, Y. Deceuninck, S. Sterk, H. J. Van Rossum, M. K. Hoffman and H. F. De Brabander, *Food Add. Contam.*, 2005, **22**, 808.
86. S. Van Cruchten, K. De Wasch, S. Impens, P. Lobeau, I. Desmedt, P. Simoens and H. F. De Brabander, *Vlaams Diergeneeskundig Tijdschrift*, 2002, **71**, 411.
87. G. Debruyckere and C. Van Peteghem, *J. Chrom. Biomed. Appl.*, 1991, **564**, 393.
88. M. Roig, J. Segura and R. Ventura, *Anal. Chim. Acta*, 2007, **586**, 184.
89. W. M. Garrett, D. J. Hoover, C. H. Shackleton and L. D. Anderson, *Endocrinology*, 1991, **129**, 2941.
90. H. F. De Brabander, Personal Communication, 2007.
91. S. B. Mahato and S. Garai, *Steroids*, 1997, **62**, 332.
92. K. Verheyden, H. Noppe, V. Mortier, J. Vercruysse, E. Claerebout, F. Van Immerseel, C. R. Janssen and H. F. De Brabander, *Anal. Chim. Acta*, 2007, **586**, 163.
93. T. Verslycke, K. De Wasch, H. F. De Brabander and C. R. Janssen, *Gen. Comp. Endocrinol.*, 2002, **126**, 190.
94. Y. S. Song, C. Jin and E. H. Park, *Archives of Pharmacal Research (Seoul)*, 2000, **23**, 599.
95. S. Poelmans, K. De Wasch, H. Noppe, N. Van Hoof, M. Van de wiele, D. Courtheyn, W. Gillis, P. Vanthemsche, C. R. Janssen and H. F. De Brabander, *Food Addit. Contam.*, 2005, **22**, 798.
96. E. N. M. Ho, K. C. H. Yiu, F. P. W. Tang, L. Dehennin, P. Plou, Y. Bonnaire and T. S. M. Wan, *J. Chrom. B*, 2004, **808**, 287.
97. C. J. M. Arts, R. Schilt, M. Schreurs and L. A. van Ginkel, *Proceedings EuroResidue III*, 1996, **212**.
98. G. Pompa, F. Arioli, M. L. Fracchiolla, C. A. S. Rossi, A. L. Bassini, S. Stella and P. A. Biondi, *Food Addit. Contam.*, 2006, **23**, 126.
99. R. Draisci, L. Palleschi, E. Ferretti, L. Lucentini and F. delli Quadri, *J. Chrom. B*, 2003, **789**, 219.
100. European Commission, Outcome of the experts meeting on the control of Boldenone in veal calves, Brussels, 30th September 2003.
101. S. Poelmans, K. De Wasch, H. Noppe, N. Van Hoof, S. Van Cruchten, B. Le Bizec, Y. Deceuninck, S. Sterk, H. J. Van Rossum, M. K. Hoffman and H. F. De Brabander, *Food Addit. Contam.*, 2005, **22**, 808.
102. S. Hartmann, M. Lacorn and H. Steinhart, *Food Chem.*, 1998, **62**, 7.
103. D. Arnold, in *Residues of Some Veterinary Drugs in Animals and Foods*, 52nd meeeting JECFA, FAO Food and Nutrition Paper 41/12, 2000.
104. D. M. Henricks and A. K. Torrence, *Journal of AOAC International*, 1978, **61**, 1280.
105. S. Biddle *et al.*, Unpublished studies on the natural occurrence of androgens and estrogens in bovine plasma, urine and bile and the effect of exogenous steroid administration on these profiles (HFL study HFL086), 2003.

106. R. Sartori, G. J. M. Rosa and M. C. Wiltbank, *J. Dairy Sci.*, 2002, **85**, 2813.
107. M. C. Lopez-Diaz, M. C. Carro, C. Cadorniga, P. Diez-Banos and M. Mezo, *Theriogenology*, 1998, **50**, 587.
108. M. D. Fray, G. E. Mann, E. C. Bleach, P. G. Knight, M. C. Clarke and B. Charleston, *Reproduction*, 2002, **123**, 281.
109. G. Gabai, L. Marinelli, C. Simontacchi and G. G. Bono, *Steroids*, 2004, **69**, 121.
110. H. J. D. Rosa and M. J. Bryant, *Small Ruminant Research*, 2003, **48**, 155.
111. P. Yuthasastrakosol, W. M. Palmer and B. E. Howland, *Reproduction*, 1975, **43**, 57.
112. R. Claus, S. Hausler and M. Lacorn, *Food Chem. Toxicol.*, 2007, **45**, 225.
113. R. Claus, D. Schopper and H.-G. Wagner, *J. Steroid Biochem.*, 1983, **19**, 725.
114. R. Claus, C. Hoang-Vu, F. Ellendorff, H. D. Meyer, D. Schopper and U. Weiler, *J. Steroid Biochem.*, 1987, **27**, 331.
115. C. G. Gravance, P. J. Casey and M. J. Erpino, *Horm. Behav.*, 1996, **30**, 22.
116. M. Hartwig, S. Hartmann and H. Steinhardt, *Zeitschrift fuer Lebensmittel Untersuchung und – Forschung A.*, 1997, **205**, 5.
117. D. H. Albert, V. V. Prasad and S. Lieberman, *Endocrinology*, 1982, **111**, 17.
118. C. H. Son, H. G. Kang and S. H. Kim, *J. Vet. Med. Sci.*, 2001, **63**, 1287.
119. I. J. Clarke and B. A. Henry, *Rev. Reprod.*, 1999, **4**, 48.
120. H. Kamada and K. Hodate, *J. Vet. Med. Sci.*, 1998, **60**, 133.
121. D. J. Carroll, M. J. Jerred, R. R. Grummer, D. K. Combs, R. A. Pierson and E. R. Hauser, *J. Dairy Sci.*, 1990, **73**, 2855.
122. S. Fritsche and H. Steinhart, *J. Anim. Sci.*, 1998, **76**, 1621.
123. L. A. Bell, T. Gimenez, J. R. Diehl and P. K. Chakraborty, *Anim. Reprod. Sci.*, 1990, **22**, 325.
124. S. Fritsche, F. J. Schwartz, M. Kirchgebner, C. Augustini and H. Steinhardt, *Meat Science*, 1998, **50**, 257.
125. S. Hartmann and H. Steinhart, *J. Chrom. B Biomed. Sci. Appl.*, 1997, **704**, 105.
126. P. L. Ryan, G. J. King and J. I. Raeside, *Anim. Reprod. Sci.*, 1990, **23**, 75.
127. M. Hirako, T. Takahashi and I. Domeki, *Theriogenology*, 2002, **57**, 1939.
128. M. Hirako, H. Takahashi and I. Domeki, *Dom. Anim. Endocrinol.*, 1996, **13**, 187.
129. T. Janowski, S. Zdunczyk, J. Malecki-Tepicht, W. Baranski and A. Ras, *Dom. Anim. Endocrinol.*, 2002, **23**, 125.
130. V. G. Smith, L. A. Edgerton, H. D. Hafs and E. M. Convey, *J. Anim. Sci.*, 1973, **36**, 391.
131. D. L. Hunter, R. E. Erb, R. D. Randel, H. A. Garverick, C. J. Callahan and R. B. Harrington, *J. Anim. Sci.*, 1970, **30**, 47.

132. T. P. W. Samuels, PhD thesis, University of Reading, UK, 2000.
133. H. Hamudikuwanda, G. Gallo, E. Block and B. R. Downey, *Anim. Reprod. Sci.*, 1996, **43**, 15.
134. N. Chaiyabutr, S. Preuksagorn, S. Komolvanich and S. Chanpongsang, *Asian Australas. J. Anim. Sci.*, 2000, **13**, 1359.
135. N. Takenouchi, K. Oshima, K. Shimada and M. Takahashi, *J. Vet. Med. Sci.*, 2004, **66**, 1315.
136. N. Isobe, M. Akita, T. Nakao, H. Yamashiro and H. Kubota, *Anim. Reprod. Sci.*, 2005, **90**, 211.
137. R. Bage, H. Gustafsson, B. Larsson, M. Forsberg and H. Rodriguez-Martinez, *Theriogenology*, 2002, **57**, 2257.
138. R. F. Veerkamp, B. Beerda and T. van der Lende, *Livest. Prod. Sci.*, 2003, **83**, 257.
139. D. S. Eley, W. W. Thatcher, H. H. Head, R. J. Collier and C. J. Wilcox, *J. Dairy Sci.*, 1981, **64**, 296.
140. D. S. Eley, W. W. Thatcher, H. H. Head, R. J. Collier, C. J. Wilcox and E. P. Call, *J. Dairy Sci.*, 1981, **64**, 312.
141. R. F. Veerkamp, J. K. Oldenbroek, H. J. Van Der Gaast and J. H. J. Werf, *J. Dairy Sci.*, 2000, **83**, 577.
142. B. A. Crooker and M. C. Lucy, *BSAS Occas. Pub. Fertil. High Production Dairy Cow*, 2001, **26**, 223.
143. H. G. Klemcke, H. G. Kattesh, J. L. Vallet, M. P. Roberts, W. J. McGuire and R. K. Christenson, *Biol. Reprod.*, 1998, **58**, 240.
144. C. D. Nancarrow, *Aust. J. Biol. Sci.*, 1983, **36**, 183.

CHAPTER 3

Hormone Use for Growth Promotion and National Programmes for Regulation of Hormone Use in Food-producing Animals

JACK F. KAY[a] AND JAMES D. MACNEIL[b]

[a] Department of Statistics and Modelling Science, University of Strathclyde, Livingstone Tower, Richmond Street, Glasgow, G1 1XH, Scotland;
[b] Department of Chemistry, St. Mary's University, Halifax, Nova Scotia, Canada B3H 3C3

3.1 First Reports of Hormone Use in Food-producing Animals

Oestrogen is a female sex hormone and in the 1930s was first reported to affect growth rates in both cattle and poultry.[1] With subsequent advances in chemistry, treatment of food-producing animals for growth-promotion purposes became common from the 1950s, and diethylstilboestrol (DES) was amongst the first synthetic oestrogens to be used commercially for this purpose.

Stob *et al.* reported on studies of bovine, ovine and poultry treated with oestrogens, including stilboestrol and dienoestrol, with the oestrogenic effect

RSC Food Analysis Monographs No.8
Analyses for Hormonal Substances in Food-producing Animals
Edited by Jack F. Kay

Published by the Royal Society of Chemistry, www.rsc.org

being monitored by the impact on mice being fed tissue from treated and untreated animals.[2] This study also is a benchmark for future studies on veterinary drug residues, as it contains a consideration of the impact of cooking on residues in tissues and concluded that cooking bovine muscle spiked with stilboestrol had little effect on the oestrogenic activity of residues in the cooked product. The paper ends almost prophetically with the comment that "it would not be at all surprising if oestrogens would be detectable in carcasses of non oestrogen treated animals".

3.2 Origins of the European Ban on Hormone Use in Food-producing Animals

DES was widely used for the treatment of problematic pregnancies in women to prevent miscarriages.[3] However, whilst later studies questioned the effectiveness of this treatment,[4] the use of DES had effectively become established and the use continued. In 1970, Herbst reported vaginal adenocarcinoma in adolescents[5] and shortly afterwards published a further paper reporting a strong association with *in utero* exposure to DES.[6] This finding was soon confirmed by others[7–9] and this led in 1971 to a ban by the Food & Drug Administration (FDA) in the USA on the use of DES during pregnancy.[10] Restrictions on the use of DES in pregnancy were introduced at different times in individual countries.[11] For example, while prescription use of DES to prevent miscarriage terminated in Belgium in 1965, this use continued until 1977 in France and Germany.

During the 1970s, diethylstilboestrol (DES) was used illegally in animal growth promotion and in the late 1970s cases of premature trelarche (breast development in girls below the age of eight years without other signs of sexual development) were observed in Puerto Rico.[12–14] The majority of girls affected were in the age range 6–24 months.[12] Allegations were made that the cause of the premature trelache was oestrogenic contamination of food or the environment.[12–18] Claims were also made that the cause was hormone supplements used in poultry or other meat,[14] although this has never been conclusively proven.

3.3 International Assessment of Hormone Use under Codex Alimentarius Commission and JECFA

3.3.1 Initial Reviews – Natural Hormones, Trenbolone Acetate and Zeranol

The consideration of the use of hormonal growth promotants by international organisations began with a recommendation from a Working Group on Health Aspects of Residues of Anabolics in Meat, which met at Bilthoven, the Netherlands, in November, 1981, under the auspices of the World Health

Organization Regional Office for Europe.[19] The recommendation from that meeting was that available data on the use of the anabolic agents trenbolone acetate and zeranol should be evaluated "for safety in use" by "the appropriate international body – the Joint FAO/WHO Expert Committee on Food Additives" (JECFA).[20] JECFA is an independent international expert committee, jointly administered by the World Heath Organization and the Food and Agriculture Organization of the United Nations, which first met in 1956.[21] The committee evaluates scientific information and provides risk assessments on the safety of food additives, veterinary drugs and contaminants in foods to FAO, WHO and their member states. In particular, JECFA provides expert recommendations on these substances used in the development of standards by the Codex Alimentarius Commission (CAC) and its subordinate committees as part of the development of standards and guidelines under the Joint FAO/WHO Food Standards Programme. JECFA is, however, independent of the CAC.

The 26th JECFA meeting in Rome in April, 1982, determined that insufficient information was available on residue concentrations, approved usage and methods of analysis to render a conclusion and requested additional toxicological information for consideration at a future meeting of the committee.[20] The 27th Meeting of the JECFA held in Geneva in April, 1983, received additional detailed information on trenbolone acetate and zeranol residues.[22] While the committee "provisionally accepted the use of trenbolone acetate as an anabolic agent for the production of meat for human consumption in accordance with good animal husbandry practice", a final decision was deferred until the results of a toxicological study that was being conducted in non-human primates became available for evaluation. A similar conclusion was reached for zeranol.

Subsequently, the first session of the Codex Committee on Residues of Veterinary Drugs in Foods (CCRVDF) met in Washington, DC, 27–31 October 1986.[23] In addition to beginning work on the development of guidelines for regulatory programmes for control of residues of veterinary drugs in foods, the Committee also established an initial priority list for compounds to be evaluated for the establishment of residue limits. This list included both trenbolone acetate and zeranol, which were already under review by JECFA, and the endogenous hormones oestradiol, progesterone and testosterone.

Additional studies on trenbolone acetate and zeranol were provided to the 32nd JECFA Meeting in Rome in June, 1987,[24] which established a temporary Acceptable Daily Intake (ADI) of 0–0.01 $\mu g\, kg^{-1}$ of body weight per day for trenbolone acetate, pending the receipt of additional information on bound residues and ongoing residue studies. An ADI of 0–0.5 $\mu g\, kg^{-1}$ of body weight per day was established for zeranol, based on the observed "no-hormonal-effect level of 0.05 mg/kg of bodyweight per day" observed in a study conducted with "ovariectomized female cynomolgus monkeys". Residue limits of 10 $\mu g\, kg^{-1}$ and 2 $\mu g\, kg^{-1}$, respectively, were recommended for zeranol residues in bovine muscle and liver. In addition, available data on the use of the endogenous hormones oestradiol, progesterone and testosterone were considered at this meeting. The Committee concluded that for each of these compounds, the amount of exogenous hormone "ingested in edible tissues from treated animals would not be capable of exerting a hormonal effect, and therefore any toxic effect, in human beings". Based on this assessment, the Committee considered an ADI unnecessary for "a hormone that is produced endogenously in human beings" and therefore did not propose residue limits. The 34th JECFA Meeting in Geneva in February 1989 received additional toxicological and residue data for trenbolone acetate and established an ADI of 0–0.02 $\mu g\ kg^{-1}$ of body weight per day for trenbolone acetate.[25] Maximum Residue Limits (MRLs) of 2 $\mu g\ kg^{-1}$ for β-trenbolone in bovine muscle and 10 $\mu g\ kg^{-1}$ for α\-trenbolone in bovine liver were recommended for consideration by the CCRVDF.

The recommendations from the 32nd JECFA Meeting were considered at Step 3 of the 8-step Codex Alimentarius process by the 2nd Session of the CCRVDF in 1987[26]and advanced to Step 5 at the 3rd Session of CCRVDF in 1988.[27] The recommendations for trenbolone acetate, however, were held at Step 4 pending further evaluation by JECFA. During discussions of residue limits for hormones at the 2nd Session of the CCRVDF, representatives of the EEC, which at that time held "observer" status, informed the Committee that the use of hormones for growth promotion would be banned within the EEC as of 1 January 1988, and that therefore EEC member states could not support the establishment of residue limits for such compounds and uses. Specifically, quoting from the Committee report:

"The Observer from the EEC stated, in reference to the conclusions of JECFA regarding hormones, that the Community had specific legislation regarding the use of hormones and that the European Community and the Member States were bound by this legislation. The European consumer was opposed to the use of hormones for fattening and demanded meat from animals which have not been so treated. The response of the Community to this consumer demand as regards to the food they eat and the enforcement they expect had been to prohibit the use of these compounds as anabolic agents. These included any substances having oestrogenic, androgenic or gestagenic action and thyreostatic substances. In consequence neither the EEC or its Member States would be able to accept residue levels for these substances when used for fattening, nor animals or the meat and meat products from animals so treated."[26]

This position was repeated at the 3rd Session and at subsequent meetings of the CCRVDF. However, at the 4th Session of the CCRVDF, held in 1989, the recommendations for residues for oestradiol, progesterone, testosterone and zeranol were advanced to Step 8 for final approval by the Codex Alimentarius Commission, while the recommendations for trenbolone acetate were returned to Step 3 for a further round of comment to allow consideration of the results from the 34th JECFA.[28] At the 5th Session of CCRVDF in 1990, the recommendations for MRLs of 2 $\mu g\,kg^{-1}$ for β-trenbolone in bovine muscle and 10 $\mu g\,kg^{-1}$ for α-trenbolone in bovine liver were advanced to Step 5.[29]

The 19th Session of the Codex Alimentarius Commission in July, 1991, discussed the recommendations from the CCRVDF concerning residues of oestradiol, progesterone and testosterone and the recommended MRLs for zeranol, but was unable to achieve a consensus on the issue, which was strongly opposed by EEC representatives.[30] The 6th Session of the CCRVDF, which met in October, 1991, recommended advancement of the proposed MRLs for trenbolone acetate to Step 8, again with opposition from EEC representatives, and also recommended that the issues regarding the other hormones referred back by the CAC should be referred to the Codex Committee on General Principles (CCGP).[31] Further discussion of the issue occurred at the 7th Session of the CCRVDF in October, 1992, which was informed that the hormone issue had been referred to the 10th Session of the Codex Committee on General Principles by the 39th Session of the Executive Committee of the CAC.[32] The continued delay in establishment of Codex standards based on the JECFA recommendations led to a statement by the representative of COMISA, the industry association, that it would recommend to its members a delay in submitting data on compounds scheduled for evaluation by the "1994 JECFA Meeting until it became clear whether or not the Commission would take any action at its meeting in July 1993 on the hormones retained at Step 8".

The 10th Session of the CCGP, held in 1992, decided to have a discussion paper on the issue prepared for further consideration at its next session.[33] The 11th Session of the CCGP, in 1994, considered the discussion paper which dealt with the role of science in the setting of Codex standards and requested the Secretariat to draft revised text for inclusion in the Codex Manual of Procedures, which would be discussed at the next session.[34] In the meantime, the 20th Session of the CAC decided also to retain the recommendations of CCRVDF on MRLs for trenbolone at Step 8, pending the results of the work assigned to CCGP.[35] The 8th Session of CCRVDF in 1994 requested a rapid resolution of the hormone issue by CCGP in its meeting report.[36] Subsequently, the 21st Session of the Codex Alimentarius Commission, meeting in Rome in July, 1995, decided by secret ballot to adopt the MRLs for the five growth hormones which had been held at Step 8, despite continued opposition from the EU member states.[37] The discussion of "other legitimate factors" continued at subsequent meetings of the CCGP.

3.3.2 Beta-agonists

This was not the end of the discussions of "growth promoters" within the Codex Alimentarius system. At the request of the CCRVDF, the 40th JECFA, meeting in 1993, reviewed available data on the use of the β-adrenoreceptor agonist, ractopamine, used to improve weight gain and carcass leanness in pigs, but was unable to establish an ADI from the available studies.[38] Further information on ractopamine was submitted for consideration by the 62nd Meeting of JECFA, which established an ADI of 0–1 μg per kg of body weight and recommended MRLs for edible tissues of pigs and cattle of 10 $\mu g\ kg^{-1}$ for muscle, 40 μg kg^{-1} for liver, 90 μg kg^{-1} for kidney and 10 μg kg^{-1} for fat.[39] The 15th Session of the CCRVDF retained these recommended MRLs at Step 4 as the JECFA report was not available sufficiently in advance of the meeting to permit delegations to fully study the recommendations.[40] However, it also questioned the rounding procedures used by JECFA in establishing an ADI and requested a reconsideration of the ADI and the MRLs. The request was considered by the 66th JECFA, which re-affirmed the procedures used in establishing the ADI and the MRLs recommended at the 62nd Meeting.[41] The 16th Meeting of the CCRVDF advanced the proposed MRLs to Step 5[42] and these were advanced to be held at Step 8 at the 17th Meeting of the CCRVDF, which noted the objections of EU delegates and those of Norway and Switzerland in its report.[43]

The 18th Meeting of the CCRVDF[44] again discussed the status of ractopamine in the light of additional residues studies submitted to the meeting by China and an opinion on ractopamine prepared by the European Food Safety Authority.[45] The CCRVDF concluded that these submissions did not constitute new data for reconsideration by JECFA and confirmed the Step 8 assignment. The MRLs for ractopamine will be referred to the CAC for final adoption as Codex MRLs, with the concerns of the EU delegates, Norway and China noted.[44]

A second β-adrenoreceptor agonist, clenbuterol, was evaluated at the 47th Meeting of JECFA, which established an ADI of 0.004 μg per kg of body weight and MRLs of 0.2 μg kg^{-1} for muscle and fat, 0.6 μg kg^{-1} for liver and kidney and 0.05 μg l^{-1} for cows' milk for the tocolytic use approved in cattle in some countries, but also recommended that clenbuterol "should not be used as a growth-enhancing agent".[46] The 11th Session of CCRVDF retained these recommendations at Step 4 due to concerns about widespread misuse of this compound.[47] The 12th Session of the CCRVDF continued to retain the MRLs recommended for edible tissue at Step 4, but advanced the proposed MRL for milk to Step 5 as "there was less likelihood of abusing clenbuterol in milking cattle".[48] The 13th Session of the CCRVDF advanced the proposed MRL for clenbuterol in milk to Step 8 and advanced the draft MRLs for clenbuterol in edible tissues to Step 5,[49] while the 14th Session of CCRVDF advanced the proposed MRLs for clenbuterol to Step 8.[50] These recommendations were adopted at the 26th Session of the Codex Alimentarius Commission.[51]

3.3.3 Corticosteroids

The available data for the corticosteroid dexamethasone were considered by the 42nd,[52] 43rd[53] and 48th[54] Meetings of JECFA, based on the inclusion of this compound on the CCRVDF priority list. The 42nd JEFA established an ADI of 0–0.015 μg per kg of body weight and recommended temporary MRLs pending provision of a satisfactory analtytical method. These recommendations were advanced to Step 5 at the 10th Session of the CCRVDF.[55] When the additional information was not received, the 48th JECFA recommended that the temporary MRLs should not be extended. Information on an analytical method for dexamethasone based on LC/MS was provided for evaluation by the 50th Meeting of JECFA, which concluded that the method "did not meet the required performance criteria for the identification and quantification of incurred residues" and was therefore "not to be considered suitable for the analysis of dexamethasone residues for regulatory purposes".[56] No MRLs were recommended for further consideration by CCRVDF. However, the 11th Session of the CCRVDF decided, based on the widespread use of the drug, to retain the temporary MRLs recommended by the 42nd JECFA at Step 7.[47] Subsequently, due to the continued absence of a suitable analytical method, the 12th Session of the CCRVDF withdrew the temporary MRLs.[48]

3.3.4 Melengestrol Acetate

The synthetic progestagen, melengestrol acetate (MGA), was placed on the priority list for JECFA evaluation by the 11th Session of the CCRVDF[47] and was evaluated in 2000 at the 54th Meeting of JECFA.[57] The Committee established an ADI of 0–0.03 $\mu g\,kg^{-1}$ of body weight and recommended temporary MRLs of 2 $\mu g\,kg^{-1}$ for beef liver and 5 $\mu g\,kg^{-1}$ for beef fat, requesting additional information on analytical methods. As information on the hormonal activity of MGA metabolites was not available, they were treated as equivalent in activity to MGA in the development of the MRLs. These temporary MRLs were advanced to Step 5 at the 13th Session of the CCRVDF.[49] An LC/MS method was provided for evaluation by the 58th Meeting of JECFA, which concluded that the method was suitable for regulatory use and recommended the MRLs be made permanent.[58] The 14th Session of the CCRVDF, meeting in 2003, retained the MRLs for MGA at Step 6 and requested a further evaluation by JECFA, based on new information and new data which were to be provided.[50]

The 62nd Meeting of JECFA in 2004 evaluated studies on the hormonal potency of MGA metabolites, which demonstrated that the most active metabolite was nine times less potent than MGA, resulting in new recommendations of MRLs of 5 $\mu g\,kg^{-1}$ for beef liver and 8 $\mu g\,kg^{-1}$ for beef fat.[39] These were submitted to the 15th Session of the CCRVDF, but were not considered as the Secretariat notified the Committee that a calculation error had been discovered during editing of the JECFA report.[40] As the rules for expert committees did not permit a revision once the JECFA meeting had ended, the matter was referred back to JECFA for review at its next meeting. The 66th Meeting of JECFA, held in 2006, recommended MRLs for edible tissues of cattle of 18 $\mu g\,kg^{-1}$ for fat, 10 $\mu g\,kg^{-1}$ for liver, 2 $\mu g\,kg^{-1}$ for kidney and 1 $\mu g\,kg^{-1}$ for muscle.[41] These were retained at Step 7 by the 16th Session of the CCRVDF due to lack of consensus, with EU delegates leading the opposition to advancement of these MRLs.[42] There was further discussion at the 17th Session of the CCRVDF, which again retained the MRLs at Step 7 on the understanding that the EU would provide new studies for evaluation at the next meeting JECFA and that the MRLs would be advanced to Step 8 at the subsequent meeting of CCRVDF if JECFA re-affirmed its recommendations.[43] The 70th Meeting of JECFA, held in Geneva in October, 2008, maintained the MRLs recommended at the 66th Meeting of the Committee.[59] No new data on MGA were presented for consideration at the 18th Meeting of the CCRVDF.[45] The CCRVDF agreed to advance MGA to Step 8 of the procedure and the MRLs for MGA will be forwarded to the CAC for adoption, noting the opposition to this advancement from the EU delegates, China, Egypt, Norway and Switzerland.

3.3.5 Reassessment of Natural Hormones (Oestradiol, Progesterone, Testosterone)

In 1995, the European Commission convened a "Scientific Conference on Growth Promotion in Meat Production", held 29 November–1 December in Brussels to review the information available on hormone use in meat production.[60] The conclusions reached by this conference did not lead to any change in the regulatory approach to hormones used for growth promotion within the EU. However, following this conference, the EU funded a number of projects to further investigate the toxicology and residues associated with these compounds. The priority list of compounds for re-evaluation in the report of the 11th Session of CCRVDF, which met in 1998, includes the three natural hormones, oestradiol, progesterone and testosterone, which were placed on the priority list by the FAO/WHO Secretariat to CCRVDF to "ensure that all the latest information had been evaluated".[47] The report notes that the EU had notified the Secretariat of the ongoing studies and requested that the JECFA evaluation should be deferred until the results of these studies were available. This review was conducted by the 52nd Meeting of JECFA, which met in 1999.[61]

The Committee concluded that oestradiol has "genotoxic potential", but also noted that it is "inactivated in the gastrointestinal tract and liver" when given orally. Based on available studies on humans receiving therapeutic doses, the Committee established an ADI of 0–0.05 $\mu g\,kg^{-1}$ of body weight for oestradiol, 0–30 $\mu g\,kg^{-1}$ of body weight for progesterone and 0–2 $\mu g\,kg^{-1}$ of body weight for testosterone. However, since residues found in animals treated with growth-promoting hormones according to approved practices were "within the physiological range of concentrations of these substances in cattle", it was concluded that "there would be no need" to specify MRLs for edible tissues of cattle where these drugs were used according to good practice in the use of veterinary drugs, a terminology used to refer to the use which has received regulatory approval from a competent national authority. These recommendations were reported to the 12th Session of the CCRVDF, which decided not to consider these new recommendations as they "did not differ significantly" from the existing MRL recommendations of "unnecessary" for these compounds.[48]

3.3.6 Recombinant Porcine Somatotropin

Another synthetic version of a natural hormone, recombinant porcine somatotropin (rpST), was also placed on the priority list by CCRVDF and three recombinant analogues of pST were evaluated by the 52nd Meeting of JECFA.[61] The Committee considered that "rpST can be used in pigs without any appreciable health risk for consumers" and established an ADI of "not specified" and made an MRL recommendation of "not specified". The 12th Session of the CCRVDF advanced the recommendation for MRLs "not specified" to Step 5 "with the understanding that their further advancement was subject to the outcome of discussion on other legitimate factors" by the CCGP.[48] The 13th Session of CCRVDF advanced the MRLs to Step 8, noting that no additional scientific information had become available and that the discussion of other legitimate factors had been concluded by the Codex Alimentarius Commission.[49] The recommendations were adopted by the 26th Session of the CAC, which noted in its report the reservations expressed by representatives of the EU.[51]

3.3.7 The Role of Science and Other Factors in Risk Analysis within the Codex Process

The 16th Session of the CCGP forwarded recommendations to the CAC concerning "the role of science and other factors in risk analysis".[62] These proposals were adopted at the 24th Session of the CAC and included a footnote concerning the relationship to WTO principles and the provisions of the SPS and TBT agreements.[63] These principles are now included in the Codex Manual of Procedures in an Appendix titled "Statements of principle concerning the role of science in the Codex decision-making process and the extent to which other factors are taken into account".[64]

The discussions within the Codex Alimentarius bodies have reflected the divergent views of EU regulatory officials and those in other countries on the appropriateness of the use of growth-promoting substances in food animal production and have been referenced and further argued before several panels established by the World Trade Organization, beginning in 1996, to assess whether the EU regulations pertaining to hormones used in growth promotion are based on legitimate scientific concerns or constitute a technical barrier to trade. Those interested in the details of the arguments advanced at these panels and the decisions taken can access the reports posted on the WTO website.[65]

3.3.8 EU Regulations Banning Hormone Use

The reports in the late 1970s that DES was used illegally in animal growth promotion and may have been implicated in premature trelarche[12–18] triggered alarm in European consumers over the potential for adverse effects in humans if hormones continued to be used in livestock production. A subsequent call for a boycott of veal by European consumer organisations had a significant adverse effect on the market.

The toxicology of hormonal growth promoters is dealt with in greater detail in Chapter 1 prepared by Professor Levy so this issue is only briefly summarised here. In September 1980, the European Community Council of Ministers adopted a declaration proposing a ban on the use of oestrogen in food-producing animals and supporting measures to harmonise legislation on veterinary medicines and animal rearing. This resulted in significant regulatory activity within Europe and which is summarised below.

- In October 1980, the European Commission proposed a ban on the use of all hormones in livestock production but excluded their use for therapeutic purposes.
- In January 1981 the Commission presented expanded proposals which permitted the controlled use of three natural hormone products for therapeutic and zootechnical purposes. These proposals also set out control measures for the production and handling of these three hormone products and proposals for testing for residues of these hormones in animals.
- On 13 February 1981, the European Parliament debated and adopted the Nielsen Report,[66] which approved the Commission's proposals, and the EC Economic and Social Committee endorsed the proposals in February 1981.
- On 31 July 1981, the EC Council of Ministers adopted Directive 81/602/EEC which banned hormones in livestock production.[67] This Directive addressed the use of five of the six hormones at issue, and the Commission was requested by the Council to provide a report on the scientific assessment of the harmful health effects of these five hormones when used for growth promotion. The Lamming Report was prepared by a scientific group of experts and presented to the Council.[68] It concluded that most of the hormones would not present any harmful health effects when used under appropriate conditions as growth promoters in animals, but that control programmes and monitoring systems for the appropriate use of these hormones were essential, and that additional scientific investigations were necessary to assess more fully the health effects of some of the hormones.
- On 31 December 1985, Directive 85/649/EEC was adopted.[69]. This banned the use of all of the six hormones for growth-promotion purposes and established provisions for the use of these hormones for therapeutic purposes.
- On 16 March 1988, Directive 88/146/EEC replaced Directive 85/649/EEC and the ban now applied to imported meat and products produced with the hormones in question outside the European Community.[70]
- Council Directive 96/22/EC of 29 April 1996 repealed earlier Directives and prohibited the use in stock farming of certain substances having a hormonal or thyrostatic action or beta agonists.[71]
- Directive 2003/74/EC implementing the WTO ruling, entered into force on 14 October 2003.[72] This legislation amended Directive 96/22/EC and

confirmed the prohibition of substances having a hormonal action for growth promotion in farm animals. Moreover, it reduced the circumstances under which 17β-oestradiol may be administered to food-producing animals for purposes other than growth promotion. Only three uses remained permissible, on a transitional basis and under strict veterinary control:
  - treatment for animal welfare reasons of foetus maceration and/or mummification;
  - pyometra in cattle; and
  - oestrus induction in cattle, horses, sheep and goats. This use was phased out by September 2006.
- In June 2006, the European Commission presented further proposals for a Directive and this has now been published as Directive 2008/97/EC.[73] This bans the remaining uses of oestradiol-related substances in food-producing animals whilst permitting their use in pet animals. The Directive also requires the European Commission to seek additional information on other hormonal substances, including the naturally occurring steroids testosterone and progesterone and the synthetic compounds trenbolone acetate, zeranol and melengesterol acetate.

3.4 Approved Use of Hormones for Growth Promotion in North America

3.4.1 Regulatory Approval Process

The approval of any use of veterinary drugs, including the use of growth-promoting hormones, is regulated by federal health authorities in both the United States of America and Canada. Information on these competent authorities, the Centre for Veterinary Medicine of the US Food & Drug Administration, the authorising agency in the United States of America,[74] and the Veterinary Drugs Directorate, Health Canada, which performs this role in Canada,[75] is available from their respective websites. In Canada, for example, there are six key steps identified in the approval of a drug for use in food animals.[76] These include:

- Metabolism and distribution studies in target food animals to determine fate and excretion pathways of the administered drug and metabolites;
- Comparative metabolism studies in laboratory animals;
- Evaluation of toxicity and carcinogenicity studies to establish an Acceptable Daily Intake (ADI);
- Establishment of Safe Total Residue for consumption and derivation of Maximum Residue Limits for specific tissues and animal-derived foods, such as milk and honey;
- Evaluation of available analytical methodologies for monitoring of compliance; and

- Establishment of withdrawal or witholding periods following administration.

The review process is both methodical and time-consuming. Once the review and approval for use have been completed, there remains a final step of consultative review with stakeholders pending final promulgation of the Maximum Residue Limits into the Canadian Food and Drugs Act[77] and Regulations.[78] During this consultation period, the approved drug is available for use, subject to the control measure of "Administrative Maximum Residue Limits", or "AMRLs". These are somewhat analogous to the approval of "temporary MRLs" in the Codex system[79] or the "provisional MRLs" which may be authorised in the EU,[80] in that they are the MRLs recommended following the product review and become final, in this case, after the completion of the consultative period.

There are six hormonal growth promoters approved in Canada and the United States for use in beef cattle. These include the three natural, or endogeneous, hormones, progesterone, testosterone and oestradiol-17β, plus three synthetic hormones, trenbolone acetate (TBA), zeranol and melengestrol acetate (MGA). The USFDA has not established tolerances for the natural hormones, but has set excess increments over the normal concentrations found in untreated animals which are considered safe for consumers. The established excess increments for oestradiol in treated animals are 120 pg g^{-1} in muscle, 480 pg g^{-1} in fat, 360 pg g^{-1} in kidney and 240 pg g^{-1} in liver[81]. The excess increments for testosterone are 0.64 ng g^{-1} in beef muscle, 1.3 ng g^{-1} in beef liver, 1.9 ng g^{-1} in beef kidney and 2.6 ng g^{-1} in beef fat,[82], while excess increments for progesterone are 3 ng g^{-1} in muscle, 6 ng g^{-1} in liver, 9 ng g^{-1} in kidney and 12 ng g^{-1} in fat.[83]

Trenbolone acetate and zeranol are anabolic agents which are administered in capsules implanted at the base of the ear (the same approved method of application for the products containing the three natural hormones). Use of trenbolone is approved in the United States in feedlot cattle, with zero withdrawal,[84] and it is considered that a tolerance is not required.[85] Zeranol is also used in cattle with zero withdrawal and also is not considered to require the establishment of a tolerance in cattle, but a tolerance of 20 ng g^{-1} has been established for the approved use of zeranol implants in sheep in the United States of America.[86]

Melengestrol acetate, a synthetic progestagen, is orally active and is administered through the feed to promote growth and to suppress oestrus in female beef cattle being fed for slaughter,[87] with a tolerance of 25 ng g^{-1} in bovine fat.[88] An additional synthetic product, altrenogest, has been approved for synchronisation of oestrus in gilts in the USA, with tolerances of 4 ng g^{-1} in liver and 1 ng g^{-1} in muscle, measured as parent compound.[89] Use of this drug is approved for a period of only 14 days, followed by a 21-day withdrawal period prior to slaughter. Approval for this use of altrenogest has also been given in Canada, with the publication of Administrative Maximum Residue Limits (AMRLs) prior to final promulgation.[90]

The use of diethylstilbestrol and related stilbenes in food-producing animals is banned in Canada, as is the use of clenbuterol.[91] The use of "exogenous oestrogenic substances" in poultry is also on the same list of substances banned in Canada.

3.4.2 Controls to Prevent Implants Entering the Food Chain

Both Canada and the United States of America are federal states, with a constitutional division of powers between federal and state authorities in the USA and similar division of powers between the federal and provincial governments in Canada. International relations and regulation of national and international trade and commerce are federal responsibilities in both countries. In practice, this means that meat produced for a market across a state/provincial boundary or a national boundary falls under the federal inspection systems. Inspection systems also exist at state and provincial levels for products produced and sold within the state or provincial boundaries.

Measures are taken to ensure that implants from slaughtered animals do not enter the food chain. For example, in federally inspected slaughter plants in Canada, where all meat destined for export or for inter-provincial shipment in Canada is processed, animals are delivered to the slaughter plant with a number of plastic and metal tags in their ears, including the unique animal identification, currently on a tag with a barcode. Before slaughter, the barcode identifying the animal is scanned into the system; the animal is then killed and hung up on a chain, where the barcode is read identifying that carcass with that particular hook on the process line. The animal is bled and then the ears are removed, leaving the hide still intact. The ID tags are removed from the ears and placed in a Canadian Food Inspection Agency plastic bag which remains with the carcass. Ears with any remaining tags are discarded into a waste container for disposal, usually by landfill. Other parts of the animal considered inedible under Canadian regulations, such as the hide, are removed and segregated in a part of the plant for inedible materials.

After removal of the inedible portions, the carcass progresses through the part of the plant where further processing occurs. The "inedible" and "edible" product areas of the plant are segregated, so staff who are working in the "inedible" area do not enter the "edible" area. There are also separate air-handling systems for the two areas.

There are additional controls to prevent rendering of ears to prevent their inclusion in a rendered product which potentially could be mixed into animal feed. The plastic tags remaining in the ears, such as visibly numbered tags that may be used in the feedlot, cause problems in the rendering process. In addition, the hair on the ears does not render, so rendering plants will not accept either ears or the ends of tails. It is therefore not only a question of having an effective inspection process to ensure the ears are removed from the food chain, but the rendering industry does not want to receive them for the very practical reasons just provided.

In a federally registered plant, animals can only be killed when the federal inspection staff, including one or more veterinarians and supported by trained inspectors, are present. The federal inspectors have the legal authority to put any area of the plant under detention, including stopping production and removing product from the food chain. A plant with a history of infractions can lose its federal registration, which effectively will put it out of business, as it cannot serve either the inter-provincial market in Canada or export markets. To meet CFIA requirements, plants must implement a HACCP (Hazard Analysis & Critical Control Point) plan, which must be approved by CFIA and which is under continuing audit from the CFIA inspection staff on the site.

With the exception of meat destined for markets in the EU, all meat, whether for domestic consumption or export, is subject to the same inspection standards. For shipments to the EU, a CFIA veterinarian provides an additional level of *ante mortem* inspection to meet EU requirements. Only animals from a "hormone free" programme, which includes additional on-farm inspection, record-keeping, veterinary oversight and residue testing, are eligible for shipment to the EU. At the time of preparation of this book, EU inspection has provided no evidence that meat from that programme contains any hormone residues in violation of EU regulations.

3.4.3 Re-implanting and Multiple Implants

In Canada, cattle brought into a feedlot to receive a high protein diet will usually receive an implant on arrival. These animals will typically be in the feedlot for about 100 days, but the implant will be effective for approximately 45–60 days. Younger animals coming into a feedlot for a grower period on a lower protein diet will also be implanted on entry, and then re-implanted when they achieve the weight at which they are put on the high-protein ration for finishing. The second implantation occurs at the point where the first implant is losing effectiveness and the finishing period again will typically be about 100 days after the second implant.

There have been several studies conducted on the effects of multiple implants on residue concentrations. Daxenberger and co-authors investigated whether the use of multiple implants, particularly at non-approved sites, could lead to higher concentrations in muscle tissues and found that there could be considerable variation in residue concentrations in proximity to such sites.[92] However, in a more recent study using trenbolone implants, where 3–4 implants were placed at the approved site in the ear, no residues which exceeded the Codex Alimentarius MRL of 2 $\mu g\,kg^{-1}$ for muscle were found in muscle samples, although there was evidence of considerable variability between residues found in individual animals, attributed to differences in individual metabolism.[93] Regulatory programmes in Canada and the United States of America have not detected any evidence that either mis-implanting or multiple implants are a common occurrence for the approved uses of hormones in feedlot cattle.

3.5 Statistical Approaches to Residue Control

The following section looks at the National Residue Control Plans in a number of producer regions and individual countries. It is not comprehensive as many countries do not publish information on either their plans or the results of their plans.

3.5.1 Europe

In the European Union, Member States are required to undertake targeted surveillance under Directive 96/23/EC for residues of veterinary medicines and a range of other substances in animals and animal products destined for the human food supply.[94] This includes sampling red meat, poultry meat, farmed fish, milk, eggs, honey and wild and farmed game. The range of substances includes surveillance for residues of hormones, which are prohibited for growth promotion purposes in the EU by Directive 96/22/EC.[71]

Each Member State must submit a residues surveillance programme to the European Commission for consideration and agreement annually. The consolidated results of the national testing plans are reported annually by the European Commission and the most recent results for the year 2007 are published on the Europa website.[95]

Directive 96/23/EC sets down the frequencies (based on annual production data) and extent of sampling and the groups of substances to be controlled for each individual commodity. Commission Decision 97/747/EC sets out the sampling requirements for milk, eggs, honey, rabbits and game meat,[96] whilst Commission Decision 98/179/EC prescribes detailed rules for the official sampling procedures and treatment of samples prior to receipt in the analytical laboratory.[97]

The table below provides the basis from the above Directives on which national residue surveillance plans are prepared. However, it is important to note that the EU requires equivalent testing for residues in countries exporting to the EU and the requirements for such testing are given in Directive 96/23/EC. The import of animals and their products treated with growth-promoting hormones, stilbenes and beta-agonists is prohibited in EU Member States under Directive 96/22/EC. Residue test plans submitted by countries in which these substances can legitimately be used will only be accepted by the European Commission if there is a "split system" in place, which assures that animals and produce intended for the EU market have never been treated at any time during their rearing. Table 3.1

Analyses for residues of veterinary medicines and other substances in samples collected under Directive 96/23/EC must be conducted in laboratories accredited to ISO 17025 standard.[98] Commission Decision 2002/657/EC lays down the performance criteria which must be met by laboratories undertaking these analyses.[99]

Whilst all EU Member States are required to submit their results to the European Commission annually few countries publish their data as a matter of

Table 3.1 Basis for calculating number of samples for residue testing in EU Member State programmes and in countries exporting to the EU.

Species	*Commodity*	*Frequency*
Bovine	Meat	0.4% of the animals slaughtered the previous year
Bovine/Ovine/Caprine	Milk	One per 15,000 tonnes of annual production – minimum 300 samples
Porcine	Meat	0.05% of the animals slaughtered the previous year
Caprine, ovine	Meat	0.05% of the animals slaughtered the previous year
Equine	Meat	No frequency or minimum number of samples established
Poultry	Meat	One per 200 tonnes of annual production (deadweight)
	Eggs	One per 1000 tonnes of annual production for human consumption – minimum 200 samples
Rabbit	Meat	10 per 300 tonnes of annual production (deadweight) for the first 3000 tonnes + 1 sample for every 300 tonnes thereafter
Farmed and wild game	Meat	At least 100 samples
Farmed fin fish	Meat	One per 100 tonnes of annual production (deadweight)
Bees	Honey	10 per 300 tonnes of annual production for human consumption for the first 3000 tonnes + 1 sample for every 300 tonnes thereafter

routine. Results for the UK are updated monthly[100] and annually[101] on the Internet.

Results exceeding legal residue limits generate additional follow-up actions, which typically include on-farm investigations and additional directed sampling, as well as removal of the affected produce from the food supply. Legal proceedings can also be instigated in such cases so every sample integrity is paramount and cannot be compromised.

In addition to sampling production in their own countries, EU Member States are required to check consignments of imported food of animal origin from non-EU countries for residues of veterinary medicines and related substances. Samples are collected at Border Inspection Posts – the point of entry into the EU. The requirements for this testing are given in Commission Regulation (EC) No 136/2004.[102]

Consignments of food which contain veterinary medicine residues in excess of Community Maximum Residue Limits or residues of substances which do not have a Community MRL or ML may not be legally placed on the EU market and are subject to rejection. If a particular residue problem is identified, the European Community or individual Member States may reinforce checks at the point of import (see Article 24 of Directive 97/78/EC).[103] Whilst reasonable efforts are made to avoid trade disruption this cannot always be avoided and import bans can be imposed pending a satisfactory resolution of the problem in the exporting country.

To assist in the detection and dissemination of information on residue-related issues with imports into the EU, the Rapid Alert System for Food and Feed (RASFF) has been in place since 1979. This was established in EU law by Regulation EC/178/2002.[104] The RASFF system gives control authorities an effective tool for exchange of information on measures taken to ensure food safety. If a RASFF notification is issued, the Commission must inform a third country:

- if it is known that a product subject to an alert notification has been exported to that country;

or

- when a product originating from that country has been the subject of a notification, so as to allow it to take corrective measures and thus avoid repetition of the problem.

Weekly updates on incidents are published by the European Commission.[105] In addition to the weekly updates, annual reports are available for RASFFs. The most recent annual report is for the year 2007.[106]

3.5.2 North America

Residue control programmes in Canada and the United States of America are generally designed to fit within the statistical guidelines found in the Codex Alimentarius.[107] Such a programme is applied on a non-biased sampling basis to a homogeneous population, which is considered by the regulatory authorities in both countries to food animal production subject to federal inspection. The integrated nature of the production, with large producers and large slaughter facilities, requires a standardised approach to production to meet industry needs. The Codex guideline for a sampling programme applied to a homogeneous population is based on the premise that 299 random samples collected over the course of the year provide 95% confidence that a violation rate of 1% or higher will be detected. In Canada, such testing is referred to as "monitoring"[108] and is applied to all major food animals species and classes for detection of residues associated with approved uses, as well as to the detection of residues which may result from the use of non-approved or banned substances. Results which exceed legal residue limits generate additional follow-up actions, which typically include on-farm investigations and additional directed sampling.

To investigate residues associated with compounds which are used less commonly or considered low risk, extra-label use of approved compounds in non-approved species where there are no grounds to suspect widespread usage or for initial introduction of a new test method to expand testing to include a new compound, a smaller pilot survey may be conducted to provide information on residue prevalence or provide further testing of the method on the diversity of samples which may be encountered in a national programme. A pilot survey will typically include 75–150 samples. Positive findings in a pilot survey will result in more intensive monitoring. This may include directed, or surveillance, sample testing. Directed sampling is used when there is reason to suspect a pattern of misuse in a particular geographical region or type of production. Such testing may lead to product detention pending risk assessment or to product condemnation when banned substances or residues in excess of legal limits are detected.

Other terms used in the Canadian programme include "blitzes", which are sampling programmes conducted over a short time period, possibly in a limited region, to obtain a clear understanding of a current situation. A blitz will typically last for a period of only about six weeks, but may involve sampling of all herds presented for slaughter during that time. "Legal" sampling is conducted to support anticipated legal action.

Compounds are selected for inclusion in the national chemical residue monitoring programme (NCRMP) based on risk assessment, again following the general principles found in the Codex Guideline.[107] Compounds are categorised according to the consumer risk associated with exposure to residues and to the probability that such an exposure could occur. Resources are then directed to the development and implementation of methods which will detect residues of the high risk and moderate risk compounds, with testing for lower risk compounds included whenever feasible in multi-residue methods or included on a rotational basis in the residue programme. Residues of hormonal compounds, or other compounds which may be used for growth promotion, such as β-agonists and thyreostats, fall within the categories included in the annual residue control programme in Canada.

The last annual report posted on the Canadian Food Inspection Agency website is for the 2004/2005 fiscal year, including summary information from reports back to the 2000/2001 fiscal year.[108] Reports for preceding years were printed and publicly available, but have not been placed on the CFIA website. Reports for more recent years are available to auditors of the residue control programme and may be obtained through Access to Information legislation, but have not been formatted and collated for posting on the website as of the date of preparation of this chapter.

The posted information for 2001/2002 indicates that 2934 tests were conducted in that fiscal year for residues of zeranol and stilbenes (including DES), 2456 for trenbolone acetate, 919 for melengestrol acetate and related gestagens, 52 for nortestosterone, 1861 for clenbuterol, 1696 for cimaterol, 304 for multiple β-agonists and 122 for thyreostats. In addition, 837 import samples were tested for hormones. The data for 2004/2005 show a reduction in testing for zeranol/stilbenes to 1816 samples and for trenbolone to 1571, but increases to 1955 for β-agonists using the multi-residue method and 685 for thyreostats. An examination of results reported for the intervening years shows a pattern of increases and decreases which reflect response to findings in an individual year, introduction of new method capabilities and changes in profile of samples scheduled for testing in any given year, with an ongoing record of high compliance. For example, the 2004/2005 testing for zeranol and stilbenes included

372 feedlot beef, 5 "hormone-free" beef, 317 horse, 191 pig, 261 veal and 670 "other", while 2001/2002 included 419 beef, 118 "hormone-free" beef, 253 horse, 380 pork, 385 veal and 1419 "other". The category "other" may include bison, deer, poultry and other species such as emu and domesticated "wild boar". The programme has continued to evolve, with the routine introduction of methods for testosterone and progesterone in tissue[109] and oestradiol in urine,[110] as well as for hormone esters in suspected injection sites.[111]

Canadian regulations require the testing of the edible meat product for residues of drugs, so most national surveys are conducted by the collection and testing of target tissue (the tissue where residues are most likely to be encountered at a violative concentration). However, in the case of suspected non-approved usage, urine samples may also be taken as a means of identifying areas or producers where additional surveillance testing of tissues from slaughter animals may be required.

Compliance was 100% in most monitoring programmes for hormone residues. Positive findings have been for approved substances, with the exception of clenbuterol use in the 1990s, and involve violations of a nature that would not generally be viewed as posing a significant risk to consumers. The use of zeranol and trenbolone acetate as implants in feedlot cattle and of melengestrol acetate as a feed additive has been approved for several decades in Canada, but residues from approved use were so low that formal MRLs were not established. There were, predictably, occasional findings of residues of these compounds in approved species. Although these findings were below the Codex MRLs for zeranol and trenbolone and USA tolerances for melengestrol acetate, they would be reported as "non-compliance" in the Annual Reports. There were also instances of use of trenbolone in a non-approved species, veal, and more recently in bison.

The "hormone-free" programme is not part of the regular national residue monitoring programme, but is a special programme developed with the beef industry to comply with EU regulatory requirements for imported meat.[112] Animals in this programme are segregated from animals maintained under standard feedlot practices. The programme includes additional veterinary supervision to verify that hormones are not used in animals registered under this programme, with testing of urine from live animals and testing of meat at slaughter according to statistical requirements established in EU regulations.

Trenbolone acetate was first approved for use in Canada in March 1994 for implantation in the ear, 6 pellets per implant, each containing 20 mg trenbolone acetate and 4 mg oestradiol, with the use restricted to non-dairy and non-breeding bovines. Prior to this approval, there were isolated instances of illegal use, resulting in short-term surveillance sampling. An example of the effectiveness of a regulatory programme combining surveillance testing with regulatory action is provided from Canadian experience with the non-approved usage of trenbolone acetate implants in late 1985 and early 1986. In this instance, Customs inspectors identified trenbolone implants being imported illegally into Canada, leading to increased surveillance by meat inspection staff at federally inspected plants. Implants found in the brisket area of a veal calf at slaughter were identified as containing trenbolone acetate, a product then used in veal production in some European countries. The intensive surveillance programme that followed included the testing of over 2000 veal calves as the programme was extended from an initially localised area to cover all major veal production areas in Canada, and led to legal action against two companies. The results demonstrated that the problem was localised to one region and, with the intensive testing programme, the violation rate rapidly dropped to zero and remained there until the late 1990s, when there was another occurrence of use of trenbolone implants in veal animals. The detection of non-approved use of trenbolone in bison has been reported in the national residue monitoring programme, as documented by EU auditors.[113]

The use of injectable testosterone was identified in some veal in 2005, with a follow-up programme targeting animals with visible injection sites presented for slaughter, combined with more general monitoring.[114] A subsequent EU audit of the Canadian residue control programme noted that "no non-compliant results for hormone residues at injection sites" had been detected from mid-2005 until the time of the audit in 2007.[113]

Testing for thyreostatic substances, used to cause weight gain through water retention prior to slaughter, has yielded no positive findings to date which have been attributed to use of these compounds in food animal production. Traces of thiouracil ($<10\,\mu g\,kg^{-1}$) have been attributed to the presence of brassica crops in feed.[108] Such sources have been documented in a published European study.[115]

The results of the Canadian monitoring programmes as documented in Annual Reports over the past two decades demonstrate that producers use licensed products according to the approved protocols and that such usage does not result in residues that exceed established Canadian or Codex Alimentarius Commission MRLs. While the survey and surveillance results demonstrate that non-approved uses of approved products or use of non-approved products does occur, such use is discouraged by a broadly based programme that is annually reviewed and expanded in scope as new test methods become available and are suitably validated. There has been no evidence of misuse of growth-promoting steroids or related products in the major feedlot beef, pork and poultry industries or in sheep in Canada.

The USA programme is very similar to the Canadian residue control programme in concept and design, as documented in the Annual "Blue Book" produced by USDA's Food Safety and Inspection Service (FSIS). The Blue Book for 2007,[116] available on-line, is in two parts, Part 1 dealing with the sampling programme for veterinary drug residues,[117] while Part 2 contains the monitoring programme for pesticide residues.[118] The National Residue Program (NRP), initiated in 1967, provides testing of meat, meat products and eggs, both from sampling at domestic plants in the USA and from imports. As with the Canadian NCRMP, there is provision for scheduled sampling based on the same statistical model referenced in Codex guidelines,[107] so that 230 or 300 random samples provide 90% or 95% confidence, respectively, of detection of a 1% violation rate for a given residue. Inspectors also may sample any suspect animals for residue testing or additional surveillance sampling may be undertaken on suspect animal populations.

For 2007, the FSIS sampling plan for hormones used in growth promotion includes testing of 230 domestic formula-fed veal and 90 samples of "fresh veal" imports for zeranol, 230 samples of domestic formula-fed veal for trenbolone acetate and 300 domestic heifers for melengestrol acetate. No testing for stilbenes or endogenous hormones is listed in the plans, but 1320 samples (1200 domestic, 120 import) are scheduled to be tested for β-agonists and testing for thyreostats is scheduled for 300 domestic formula-fed veal and 90 fresh veal imports. Decisions on which substances to include in the annual NRP, and in which species, are determined based on the probability of exposure and the risk associated with an exposure, as explained in the Blue Book, Part 1.[117] Results of the NRP testing are published annually in the "Red Book", with the most recent report being for 2006.[119] The findings include 1 "non-formula-fed veal" positive for the β-agonist salbutamol (out of 939 veal and heifer samples tested) and 2 of the 174 "non-formula-fed veal" positive for trenbolone. None of the 323 formula-fed veal tested were found positive for trenbolone or zeranol.

In addition, a "non-hormone treated cattle" programme, or "NHTC Program", is administered by the Agricultural Marketing Service of the USDA.[120] The programme combines veterinary certification, segregation from hormone-treated animals and industry-funded residue testing as per EU requirements in accredited private laboratories.[121]

3.5.3 Australia

The National Residue Survey (NRS) in Australia dates back to 1961 and was set up to counter concerns about pesticide residue in exported meat from Australia. The NRS has expanded over the years and now includes testing for residues of agricultural chemicals, veterinary medicines and environmental contaminants in 25 animal and 25 plant commodities including 5 horticultural products.

For residues of veterinary medicines and related substances, the sampling plans are reviewed annually and must meet the requirements of both the Australian Quarantine and Inspection Service and those of international trading partners (*e.g.* the EU). Sampling collection rates are based on annual production data for the internal production or by overseas legislation for export purposes.

Further information on the NRS and a summary of the results for the year 2007/8 are published on the Internet.[122]

3.5.4 South America

MERCOSUR is an organisation which is effectively the Common Market of South America and links Argentina, Brazil, Paraguay and Uruguay.

Venezuela has applied for full membership of MERCOSUR but this has yet to be ratified. Bolivia, Chile, Colombia, Ecuador and Peru are currently associate members of this group. The current four members are the major meat-exporting countries of South America with a total bovine herd in excess of 200 million animals.

Whilst there is national legislation on residue controls in individual countries of MERCOSUR, consideration is being given to harmonising residue control plans which are drawn up to meet both internal and export needs. In general terms, three types of programmes can be undertaken:

- Random sampling (monitoring) – Sampling is undertaken to detect violations of authorised substances typically at an incidence of 1% with a statistical confidence of 95%. This requires the collection of 300 samples. However, for detection of non-authorised substances, an incidence of 0.1% is sought with 95% statistical confidence. This requires the collection of 3000 samples. If either of the above incidences is exceeded this triggers direct sampling (below). The actual number of samples collected will be related to the annual production.
- Direct sampling (surveillance) – If random sampling above detects an incidence in excess of the values of 0.1% or 1% respectively, direct targeted sampling, in addition to any requirements in a national plan, is instituted against the area or individual farm identified as the source of the problem.
- Special sampling – An importing country or market can request special sampling to be put in place by a MERCOSUR country to meet their individual import regulations. For example, the EU requires testing to demonstrate the absence of hormone treatment in animals exported to the EU.

Fernandez Suarez presents further detail on the current MERCOSUR position in the report of the Joint FAO/WHO Technical Workshop on Residues of Veterinary Drugs Without ADI/MRL held in Thailand in 2004.[123]

Results of the national residue control programmes in the MERCOSUR countries are not published but some information on their effectiveness can be gained from the reports published by the European Communities FVO following inspections in individual countries. The latest FVO report on Argentina was published in 2008.[124]

This report comments that the prohibition of the use of hormones, beta-agonists and thyrostats in growth promotion in Argentina provides confidence in the residues status of produce exported to the EU but there were some issues with, for example, sampling frequency across the year.

For Brazil, the latest FVO was also published in 2008.[125] This report states that the national residue control programme in Brazil largely meets the requirements of Commission Directive 96/23/EC.[91] However, it concludes that whilst progress has been made over recent years, the residue control system is not as effective as it could be, highlighting particularly that the scope of testing could be improved for some substance groups (including increased sampling for hormones) and that the supervision of the plan by the authorities could be improved, although recent measures adopted suggested that this would be resolved.

The latest FVO report for Uruguay was published in 2004.[126] Responsibility for the national control programme for residues in Uruguay rests with the Ministry of Livestock, Agriculture and Fisheries (MGAP). In addition to the testing for exports to the EU, there is additional testing for zeranol in bovine produce destined for the USA. The FVO considered that the implementation, supervision and follow-up on the national residue control plan were effective but recommended that further substances should be included in the plan and additional samples be analysed in particular areas.

3.5.5 People's Republic of China

The Ministry of Agriculture (MoA) and the General Administration of Quality, Supervision, Inspection and Quarantine of the People's Republic of China (AQSIQ) collaborate in the planning, co-ordination, supervision and follow-up of the annual national residues control plan. AQSIQ is responsible for all exported commodities and MoA is responsible for the control and supervision of the domestic market. MoA is also involved in the sampling and follow up of non-compliant samples which are approved under the Export Oriented System.

Further information on the national residue control plan in the People's Republic of China is available from the 2006 EU FVO report.[127] This report concluded that there is a comprehensive residues control plan in place in the People's Republic of China which is largely in line with the European Community requirements.

3.5.6 Thailand

In Thailand, the national residues control plan is based upon the requirements of Directive 96/23/EC.[71] Responsibility for this plan is shared between the Department of Livestock Development (DLD) and the Department of Fisheries (DOF). The most recent inspection by the European Union FVO on Thailand was in 2006, at which time the main exports to the EU were poultry, honey and fish.[128]

The report concluded that overall the system in Thailand was working satisfactorily but highlighted a number of issues, *e.g.* the lack of availability of some analytical methods, which needed further attention.

References

1. R. Gandhi and S. M. Snedeker, http://envirocancer.cornell.edu/Factsheet/Diet/fs37.hormones.cfm, accessed 23 March 2009.

2. M. Stob, F. N. Andrews, M. X. Zarrow and W. M. Beeson, *J. Anim. Sci.*, 1954, **13**, 138–151.
3. O. W. Smith and G. V. Smith, *N. Engl. J. Med.*, 1949, **241**, 562–568.
4. J. H. Ferguson, *Am. J. Obstet. Gynecol.*, 1953, **65**, 592–601.
5. A. L. Herbst and R. E. Scully, *Cancer*, 1970, **25**, 745–747.
6. A. L. Herbst, H. Ulfelder and D. C. Poskanzer, *N. Engl., J. Med.*, 1971, **284**, 878–881.
7. P. Greenwald, J. J. Barlow, P. C. Nasca and W. S. Burnett, *N. Engl. J. Med.*, 1971, **285**, 390–392.
8. E. C. Hill, *Am. J. Obstet. Gynecol.*, 1973, **116**, 470–484.
9. B. E. Henderson, B. D. Benton, P. T. Weaver, G. Linden and F. Nolan, *N. Engl. J. Med.*, 1973, **288**, 354–358.
10. *United States Federal Register* 36, 217: 21537–8, November 10, 1971.
11. D. Ibarreta and S. H. Swan, in Late lessons from early warnings: the precautionary principle 1896–2000, European Environment Agency, http://reports.eea.europa.eu/environmental_issue_report_2001_22/en/issue-22-part-08.pdf, accessed 23 March 2009.
12. C. A. Saenz, M. A. Toro-Sola, L. Conde and N. P. Bayonet-Rivera, *Bol. Assoc. Med. PR*, 1982, **74**, 16–18.
13. C. A. Perez, *Lancet*, 1982, **1**, 1299.
14. A. M. Bongiovanni, *J. Pediatr.*, 1983, **102**, 245–246.
15. C. A. Saenz and M. A. Toro-Sola, *Lancet*, 1983, **1**, 537.
16. R. Schoental, *Lancet*, 1983, **1**, 537.
17. C. A. Saenz, *New Engl. J. Med.*, 1984, **310**, 1741–1742.
18. C. A. Saenz, A. M. Bongiovanni and L. Conde, *J. Pediatr.*, 1985, **107**, 393–396.
19. Health aspects of residues of anabolics in meat, Copenhagen, WHO Regional Office for Europe, 1982 (EURO Reports and Studies 59).
20. Evaluation of certain food additives and contaminants, 26th Report of the Joint FAO/WHO Expert Committee on Food Additives, Technical Report Series 683, World Health Organization, Geneva, 1982.
21. Fact Sheet – What is JECFA?, FAO/WHO Joint Secretariat to JECFA, 09/02/2006, ftp://ftp.fao.org/ag/agn/jecfa/jecfa_2006-02.pdf, accessed 23 March 2009.
22. Evaluation of certain food additives and contaminants, 27th Report of the Joint FAO/WHO Expert Committee on Food Additives, Technical Report Series 696, World Health Organization, Geneva, 1983.
23. Report of the First Session of the Codex Committee on Residues of Veterinary Drugs in Foods, Alinorm 87/31, Codex Alimentarius Commission, 1987, http://www.codexalimentarius.net/download/report/206/al87_31e.pdf, accessed 23 March 2009.
24. Evaluation of certain veterinary drug residues in food, 32nd Report of the Joint FAO/WHO Expert Committee on Food Additives, Technical Report Series 763, World Health Organization, Geneva, 1988.

25. Evaluation of certain veterinary drug residues in food, 34th Report of the Joint FAO/WHO Expert Committee on Food Additives, Technical Report Series 788, World Health Organization, Geneva, 1989.
26. Report of the Second Session of the Codex Committee on Residues of Veterinary Drugs in Foods, Alinorm 89/31, Codex Alimentarius Commission, 1989, http://www.codexalimentarius.net/download/report/207/al89_31e.pdf, accessed 23 March 2009.
27. Report of the Third Session of the Codex Committee on Residues of Veterinary Drugs in Foods, Alinorm 89/31A, Codex Alimentarius Commission, 1989, http://www.codexalimentarius.net/download/report/208/al8931ae.pdf, accessed 23 March 2009.
28. Report of the Fourth Session of the Codex Committee on Residues of Veterinary Drugs in Foods, Alinorm 91/31, Codex Alimentarius Commission, 1991, http://www.codexalimentarius.net/download/report/209/al91_31e.pdf, accessed 23 March 2009.
29. Report of the Fifth Session of the Codex Committee on Residues of Veterinary Drugs in Foods, Alinorm 91/31A, Codex Alimentarius Commission, 1991, http://www.codexalimentarius.net/download/report/210/al9131be.pdf, accessed 23 March 2009.
30. Report of the Nineteenth Session of the Codex Alimentarius Commission, Alinorm 91/40, Codex Alimentarius Commission, 1991, http://www.fao.org/docrep/meeting/005/t0490e/t0490e00.htm, accessed 23 March 2009.
31. Report of the Sixth Session of the Codex Committee on Residues of Veterinary Drugs in Foods, Alinorm 93/31, Codex Alimentarius Commission, 1991, http://www.codexalimentarius.net/download/report/211/al93_31e.pdf, accessed 23 March 2009.
32. Report of the Seventh Session of the Codex Committee on Residues of Veterinary Drugs in Foods, Alinorm 93/31A, Codex Alimentarius Commission, 1991, http://www.codexalimentarius.net/download/report/212/al9331ae.pdf, accessed 23 March 2009.
33. Report of the Tenth Session of the Codex Committee on General Principles, Alinorm 93/33, Codex Alimentarius Commission, 1993, http://www.codexalimentarius.net/download/report/1/al93_33e.pdf, accessed 23 March 2009.
34. Report of the Eleventh Session of the Codex Committee on General Principles, Alinorm 95/33, Codex Alimentarius Commission, 1995, http://www.codexalimentarius.net/download/report/2/al95_33e.pdf, accessed 23 March 2009.
35. Report of the Twentieth Session of the Codex Alimentarius Commission, Alinorm 93/40, Codex Alimentarius Commission, 1993, http://www.fao.org/docrep/meeting/005/t0817e/t0817e00.htm, accessed 23 March 2009.
36. Report of the Eighth Session of the Codex Committee on Residues of Veterinary Drugs in Foods, Alinorm 95/31, Codex Alimentarius

Commission, 1995, http://www.codexalimentarius.net/download/report/213/AL95_31e.pdf accessed 23 March 2009.
37. Report of the Twenty-first Session of the Codex Alimentarius Commission, Alinorm 95/37, Codex Alimentarius Commission, 1995, http://www.fao.org/docrep/meeting/005/v7950e/v7950e00.htm, accessed 23 March 2009.
38. Evaluation of certain veterinary drug residues in food, 40th Report of the Joint FAO/WHO Expert Committee on Food Additives, Technical Report Series 832, World Health Organization, Geneva, 1993.
39. Evaluation of certain veterinary drug residues in food, 62nd Report of the Joint FAO/WHO Expert Committee on Food Additives, Technical Report Series 925, World Health Organization, Geneva, 2004.
40. Report of the Fifteenth Session of the Codex Committee on Residues of Veterinary Drugs in Foods, Alinorm 05/28/31, Codex Alimentarius Commission, 2005, http://www.codexalimentarius.net/download/report/626/al28_31e.pdf, accessed 23 March 2009.
41. Evaluation of certain veterinary drug residues in food, 66th Report of the Joint FAO/WHO Expert Committee on Food Additives, Technical Report Series 939, World Health Organization, Geneva, 2006.
42. Report of the Sixteenth Session of the Codex Committee on Residues of Veterinary Drugs in Foods, Alinorm 06/29/31, Codex Alimentarius Commission, 2006, http://www.codexalimentarius.net/download/report/659/al29_31e.pdf, accessed 23 March 2009.
43. Report of the Seventeenth Session of the Codex Committee on Residues of Veterinary Drugs in Foods, Alinorm 08/31/31, Codex Alimentarius Commission, 2008, http://www.codexalimentarius.net/download/report/685/al31_31e.pdf, accessed 23 March 2009.
44. Report of the Eighteenth Session of the Codex Committee on Residues of Veterinary Drugs in Foods, Alinorm 09/32/3, Codex Alimentarius Commission, 2009, ftp://ftp.fao.org/codex/alinorm09/al32_31e.pdf, accessed 31 May 2009.
45. *EFSA Journal*, 2009, **1041**, 1–52.
46. Evaluation of certain veterinary drug residues in food, 47th Report of the Joint FAO/WHO Expert Committee on Food Additives, Technical Report Series 876, World Health Organization, Geneva, 1998.
47. Report of the Eleventh Session of the Codex Committee on Residues of Veterinary Drugs in Foods, Alinorm 99/31, Codex Alimentarius Commission, 1999, http://www.codexalimentarius.net/download/report/216/Al99_31e.pdf, accessed 23 March 2009.
48. Report of the Twelfth Session of the Codex Committee on Residues of Veterinary Drugs in Foods, Alinorm 01/31, Codex Alimentarius Commission, 2001, http://www.codexalimentarius.net/download/report/217/Al01_31e.pdf, accessed 23 March 2009.
49. Report of the Thirteenth Session of the Codex Committee on Residues of Veterinary Drugs in Foods, Alinorm 03/31, Codex Alimentarius

Commission, 2003, http://www.codexalimentarius.net/download/report/218/al03_31e.pdf, accessed 23 March 2009.
50. Report of the Fourteenth Session of the Codex Committee on Residues of Veterinary Drugs in Foods, Alinorm 03/31A, Codex Alimentarius Commission, 2003, http://www.codexalimentarius.net/download/report/45/Al0331ae.pdf, accessed 23 March 2009.
51. Report of the Twenty-sixth Session of the Codex Alimentarius Commission, Alinorm 03/41, Codex Alimentarius Commission, 2003, http://www.codexalimentarius.net/download/report/601/al03_41e.pdf, accessed 23 March 2009.
52. Evaluation of certain veterinary drug residues in food, 42nd Report of the Joint FAO/WHO Expert Committee on Food Additives, Technical Report Series 851, World Health Organization, Geneva, 1995.
53. Evaluation of certain veterinary drug residues in food, 43rd Report of the Joint FAO/WHO Expert Committee on Food Additives, Technical Report Series 855, World Health Organization, Geneva, 1995.
54. Evaluation of certain veterinary drug residues in food, 48th Report of the Joint FAO/WHO Expert Committee on Food Additives, Technical Report Series 879, World Health Organization, Geneva, 1998.
55. Report of the Tenth Session of the Codex Committee on Residues of Veterinary Drugs in Foods, Alinorm 97/31A, Codex Alimentarius Commission, 1997, ftp://ftp.fao.org/docrep/fao/meeting/005/w3945e/w3945e.pdf, accessed 23 March 2009.
56. Evaluation of certain veterinary drug residues in food, 50th Report of the Joint FAO/WHO Expert Committee on Food Additives, Technical Report Series 888, World Health Organization, Geneva, 1999.
57. Evaluation of certain veterinary drug residues in food, 54th Report of the Joint FAO/WHO Expert Committee on Food Additives, Technical Report Series 900, World Health Organization, Geneva, 2001.
58. Evaluation of certain veterinary drug residues in food, 58th Report of the Joint FAO/WHO Expert Committee on Food Additives, Technical Report Series 911, World Health Organization, Geneva, 2002.
59. Joint FAO/WHO Expert Committee on Food Additives, Seventieth meeting (Residues of veterinary drugs), Geneva, 21–29 October 2008, Summary and Conclusions – Corrected. Issued 10 March 2009, FAO/WHO, http://www.fao.org/ag/agn/agns/jecfa/JECFA70_Summary_report_final_corr.pdf, accessed 23 March 2009.
60. EC, Proceedings of the Scientific Conference on Growth Promotion in Meat Production, European Commission Directorate-General VI Agriculture, Brussels, Belgium, November 29 – December 1, 1995, Office for Official Publications of the European Communities, 1996, ISBN 92-827-6321-8.
61. Evaluation of certain veterinary drug residues in food, 52nd Report of the Joint FAO/WHO Expert Committee on Food Additives, Technical Report Series 893, World Health Organization, Geneva, 2000.

62. Report of the Sixteenth Session of the Codex Committee on General Principles, Alinorm 01/33A, Codex Alimentarius Commission, 2001, http://www.codexalimentarius.net/download/report/89/Al0133ae.pdf, accessed 23 March 2009.
63. Report of the Twenty-fourth Session of the Codex Alimentarius Commission, Alinorm 01/41, Codex Alimentarius Commission, 2001, http://www.codexalimentarius.net/download/report/519/Al01_41e.pdf, accessed 23 March 2009.
64. Procedural Manual, 17th edition, Codex Alimentarius Commission, 2008, ftp://ftp.fao.org/codex/Publications/ProcManuals/Manual_17e.pdf, accessed 23 March 2009.
65. Index of disputes issues, World Trade Organization, 2008, http://www.wto.org/english/tratop_e/dispu_e/dispu_subjects_index_e.htm#hormones,_meat, accessed 23 March 2009.
66. WTO, SICE Foreign Trade Information System – WT/DS48/R/CAN 18 August 1997, EC Measures Concerning Meat and Meat Products (Hormones), Complaint by Canada – Report of the Panel, http://www.sice.oas.org/DISPUTE/wto/horm-can.asp, 2009, accessed 31 March 2009.
67. Official Journal of the European Community L 222, 7 August 1981, pp. 0032–0033.
68. B. Krissoff, The European ban on livestock hormones and implications for international trade, National Food Review, FindArticles.com, 15 Mar. 2009, http://findarticles.com/p/articles/mi_m3284/is_n3_v12/ai_8072077, accessed 23 March 2009.
69. Official Journal of the European Community L 382, 31 December 1985, pp. 228–231.
70. Official Journal of the European Community L 070, 16 March 1988, pp. 0016–0018.
71. Official Journal of the European Community L 125, 23 May 1996, pp. 0003–0009.
72. Official Journal of the European Union L 262, 14 October 2003, pp. 0017–0021.
73. Official Journal of the European Union L 318, 28 November 2008, pp. 0009–0011.
74. FDA, http://www.fda.gov/cvm/, 2008, accessed 23 March 2009.
75. HC, http://www.hc-sc.gc.ca/dhp-mps/vet/index-eng.php, 2008, accessed 23 March 2009.
76. HC, Setting Standards for Maximum Residue Limits (MRLs) of Veterinary Drugs Used in Food-Producing Animals, http://www.hc-sc.gc.ca/dhp-mps/vet/mrl-lmr/mrl-lmr_levels-niveaux-eng.php, 2008, accessed 23 March 2009.
77. Department of Justice Canada, Food and Drugs Act, http://laws.justice.gc.ca/en/showtdm/cs/F-27///en, 2008, accessed 23 March 2009.
78. Department of Justice Canada, Food and Drug Regulations, http://laws.justice.gc.ca/en/showtdm/cr/C.R.C.-c.870, 2008, accessed 23 March 2009.

79. FAO, Joint FAO/WHO Expert Committee on Food Additives (JECFA): Procedures for Recommending Maximum Residue Limits – Veterinary Drugs in Foods, ftp://ftp.fao.org/es/esn/jecfa/2000-06-30_JECFA_Procedures_MRLVD.pdf, 2008, accessed 23 March 2009.
80. European Medicines Agency, Status of MRL Procedures – MRL Assessments in the context of Council Regulation (EEC) No.2377/90, http://www.emea.europa.eu/pdfs/vet/mrls/076599en.pdf, 2008, accessed 23 March 2009.
81. FDA, Code of Federal Regulations, 21 CFR 556.240, http://www.accessdata.fda.gov/scripts/cdrh/cfdocs/cfcfr/CFRSearch.cfm?fr=556.240, 1991, accessed 23 March 2009.
82. FDA, Code of Federal Regulations, 21 CFR 556.710, http://www.accessdata.fda.gov/scripts/cdrh/cfdocs/cfcfr/CFRSearch.cfm?fr=556.710, 2007, accessed 23 March 2009.
83. FDA, Code of Federal Regulations, 21 CFR 556.540, http://www.accessdata.fda.gov/scripts/cdrh/cfdocs/cfcfr/CFRSearch.cfm?fr=556.540, 2007, accessed 23 March 2009.
84. FDA, Freedom of Information Summary, Supplemental New Animal Drug Application, NADA 141-043, http://www.fda.gov/cvm/FOI/141-043s100302.pdf, 2002, accessed 23 March 2009.
85. FDA, Code of Federal Regulations, 21 CFR 556.739, http://www.accessdata.fda.gov/scripts/cdrh/cfdocs/cfcfr/CFRSearch.cfm?fr=556.739, 2007, accessed 23 March 2009.
86. FDA, Code of Federal Regulations, 21 CFR 556.760, http://www.accessdata.fda.gov/scripts/cdrh/cfdocs/cfcfr/CFRSearch.cfm?fr=556.760, 2007, accessed 23 March 2009.
87. FDA, Freedom of Information Summary, Original New Animal Drug Application, NADA 039-402, http://www.fda.gov/cvm/FOI/514.htm, 1999, accessed 23 March 2009.
88. FDA, Code of Federal Regulations, 21 CFR 556.380, http://www.accessdata.fda.gov/scripts/cdrh/cfdocs/cfcfr/CFRSearch.cfm?fr=556.380, 2007, accessed 23 March 2009.
89. FDA, Freedom of Information Summary, Original New Animal Drug Application, NADA 141-222, Altrenogest (Matrix), http://www.fda.gov/cvm/FOI/141-222.pdf, 2008, accessed 23 March 2009.
90. HC, Administrative Maximum Residue Limits (AMRLS) and Maximum Residue Limits (MRLS) set by Canada, http://www.hc-sc.gc.ca/dhp-mps/vet/mrl-lmr/mrl-lmr_versus_new-nouveau-eng.php, 2008, accessed 23 March 2009.
91. HC, List of banned drugs, Health Canada, http://www.hc-sc.gc.ca/dhp-mps/vet/banned_drugs_list_interdit_medicaments-eng.php, 2005, accessed 23 March 2009.
92. A. Daxenberger, I. G. Lange, K. Meyer and H. H. D. Mayer, *J. AOAC Int.*, 2000, **83**, 809.
93. J. D. MacNeil, J. Reid and R. W. Fedeniuk, *J. AOAC Int.*, 2008, **91**, 670.

94. Official Journal of the European Community L 125, 23 May 1996, pp. 10–32.
95. http://ec.europa.eu/food/food/chemicalsafety/residues/workdoc_2007_en.pdf, accessed 23 March 2009.
96. Official Journal of the European Community L 303, 6 November 1997, pp. 12–15.
97. Official Journal of the European Community, L 065, 5 March 1998, pp. 31–34.
98. http://www.iso.org/iso/Catalogue_detail?csnumber=39883, accessed 23 March 2009.
99. Official Journal of the European Union L 221, 17 August 2002, pp. 8–36.
100. http://www.vmd.gov.uk/Publications/MAVIS/mavis.htm, accessed 23 March 2009.
101. http://www.vet-residues-committee.gov.uk/Reports/annual.htm, accessed 23 March 2009.
102. Official Journal of the European Union, L 21, 28 January 2004, pp. 11–23.
103. Official Journal of the European Community, L 24, 30 January 1998, pp. 9–30.
104. Official Journal of the European Community, L 31, 1 February 2002.
105. http://ec.europa.eu/food/food/rapidalert/archive_en.htm, accessed 23 March 2009.
106. http://ec.europa.eu/food/food/rapidalert/report2007_en.pdf, accessed 23 March 2009.
107. CAC, Codex guidelines for the establishment of a regulatory programme for control of veterinary drug residues in foods. CAC/GL 16–1993, http://www.codexalimentarius.net/download/standards/47/CXG_016e.pdf, 1995, accessed 23 March 2009.
108. CFIA, Report on Pesticides, Agricultural Chemicals, Veterinary Drugs, Environmental Pollutants and Other Impurities in Agri-Food Commodities of Animal Origin, Fiscal Year 2004/2005, http://www.inspection.gc.ca/english/fssa/microchem/resid/reside.shtml, 2008, accessed 31 March, 2009.
109. R. W. Fedeniuk, M. West, R. Gedir, M. Mizuno, C. Neiser and J. D. MacNeil, *J. AOAC. Int.*, 2006, **89**, 576.
110. C. Akre, R. Fedeniuk and J. D. MacNeil, *Analyst*, 2004, **129**, 145.
111. R. M. Costain, A. C. E. Fesser, D. McKenzie, M. Mizuno and J. D. MacNeil, *Food Addit. Contam.*, 2008, **25**, 1520–1529.
112. CFIA, Meat Hygiene Manual of Procedures, Chapter 11 – Annex R: Program for certifying freedom from hormonal growth promotants (HGPs) (Hormonal Growth Promotants and beta-agonists having an anabolic effect), http://www.inspection.gc.ca/english/anima/meavia/mmopmmhv/chap11/eu-ue/annexre.shtml, 2008, accessed 23 March 2009.
113. EU, Final report of a mission carried out in Canada from 28 May to 7 June 2007 concerning the evaluation of the control of residues and

contaminants in live animals and animal products, including controls on veterinary medicinal products. DG(SANCO)/2007-731 MR Final, Directorate F, Foreign Veterinary Office, Health and Consumer Protection Directorate-General, European Commission, http://ec.europa.eu/food/fvo/act_getPDF.cfm?PDF_ID=6196, 2007, accessed 23 March 2009.

114. R. M. Costain, A. C. E. Fesser, D. McKenzie, M. Mizuno and J. D. MacNeil, *Food Addit. Contam.*, 2008, **25**, 1520–1529.
115. G. Pinel, S. Mathieu, N. Cesbron, D. Maume, H. F. De Brabander, F. Andre and B. Le Bizec, *Food Addit. Contam.*, 2006, **23**, 974.
116. FSIS, 2007 FSIS National Residue Program Scheduled Sampling Plans – "Blue Book", Food Safety & Inspection Service, United States Department of Agriculture, http://www.fsis.usda.gov/Science/2007_Blue_Book/index.asp, 2007, accessed 23 March 2009.
117. FSIS, 2007 FSIS National Residue Program Scheduled Sampling Plans – "Blue Book", Part 1, Food Safety & Inspection Service, United States Department of Agriculture, http://www.fsis.usda.gov/PDF/2007_Blue_Book_Part1.pdf, 2007, accessed 23 March 2009.
118. FSIS, 2007 FSIS National Residue Program Scheduled Sampling Plans – "Blue Book", Part 2, Food Safety & Inspection Service, United States Department of Agriculture, http://www.fsis.usda.gov/PDF/2007_Blue_Book_Part2.pdf, 2007, accessed 23 March 2009.
119. FSIS, 2006 FSIS National Residue Program Data, Food Safety & Inspection Service, United States Department of Agriculture, http://www.fsis.usda.gov/PDF/2006_Red_Book.pdf, 2006, accessed 23 March 2009.
120. AMS, USDA Export Verification (EV) Program Specified Product Requirements Non-Hormone Treated Cattle (NHTC) for the European Union, http://www.ams.usda.gov/AMSv1.0/getfile?dDocName=STELDEV3103506, 2006, accessed 23 March 2009.
121. AMS, Grading, Certification and Verification, LS ARC Questions and Answers – NHTC, Agricultural Marketing Service, USDA, http://www.ams.usda.gov/AMSv1.0/ams.fetchTemplateData.do?template=TemplateN&navID=NHTC&rightNav1=NHTC&topNav=&leftNav=FairTradingRegulations&page=LSAuditingServicesARCQuestions-andAnswersPage4 &resultType=&acct=audrevcom, 2008, accessed 23 March 2009.
122. http://www.daff.gov.au/agriculture-food/nrs/publications/annual-reports, accessed 30 March 2009.
123. ftp://ftp.fao.org/docrep/fao/008/y5723e/y5723e00.pdf, accessed 23 March 2009.
124. http://ec.europa.eu/food/fvo/rep_details_en.cfm?rep_id=2037, accessed 23 March 2009.
125. http://ec.europa.eu/food/fvo/rep_details_en.cfm?rep_id=2018, accessed 23 March 2009.

126. http://ec.europa.eu/food/fvo/rep_details_en.cfm?rep_id=1173, accessed 23 March 2009.
127. http://ec.europa.eu/food/fvo/rep_details_en.cfm?rep_id=1535, accessed 23 March 2009.
128. http://ec.europa.eu/food/fvo/rep_details_en.cfm?rep_id=1642, accessed 23 March 2009.

CHAPTER 4

Current Analytical Methods Used for the Detection of Hormone Residues

MATTHEW SHARMAN,[a] LEEN VAN GINKEL[b] AND JAMES D. MACNEIL[c]

[a] Central Science Laboratory, York, UK; [b] RIVM, Bilthoven, The Netherlands; [c] Department of Chemistry, St. Mary's University, Halifax, Nova Scotia, Canada B3H 3C3

4.1 Introduction

The effective detection of "illegal" hormone residues is a demanding area of science that requires expertise and information from a wide range of disciplines such as metabolomics, proteomics, histopathology and analytical chemistry/biochemistry.

Although many of the analytical methods presented in this chapter are relatively well established it should be remembered that a detailed knowledge of the type of sample under test will be required if the analyst or Competent Authority is to correctly determine whether a hormonal substance has been employed. For example, the factors to be considered during final data analysis could include:

a) Age and sex of the animal
b) Species and breed of animal
c) Stage of gestation
d) Type of sample *e.g.* urine, feed, muscle

RSC Food Analysis Monographs No.8
Analyses for Hormonal Substances in Food-producing Animals
Edited by Jack F. Kay

Published by the Royal Society of Chemistry, www.rsc.org

e) Type of hormonal treatments permitted within a particular country/ region
f) Storage history of the sample and therefore analyte stability
g) Other external factors *e.g.* the presence of mycotoxins

Recent examples of the importance of these data interpretation issues for natural steroids include (i) *β*-boldenone and (ii) 17α,19-nortestosterone. In the first case, research has shown that free *β*-boldenone is endogenous in intact male horses, and also can be of natural origin in bovines.[1,2] In contrast, should the conjugated form be detected in bovine urine, this is usually regarded as proof of an illegal treatment.[3] In the second case of 17α,19-nortestosterone, this compound is known to be endogenous in very young calves and pregnant cows,[4] and was originally thought not to occur in male bovines.[5] Recent studies, however, have suggested that the presence of 17α,19-nortestosterone in male bovines may be produced *in vivo* after injury and, in this special case, residues may be of "natural" origin.[6] Thus while the methods reviewed in this chapter are all capable of providing high-quality data, it is often the skill and knowledge of the expert reviewing these data that is key to a successful outcome.

Screening methods for hormones are wide ranging, and include histopathological tests, Enzyme Linked Immuno Sorbent Assays (ELISAs) and Reporter Gene Assays (RGAs). Depending upon the chosen assay format these tests are capable of providing targeted screening of either individual or groups of specific substances or relatively non-specific (or "general") screening. In the EU, screening assays are designed to meet the requirements of Commission Decision 2002/657/EC,[7] which defines screening methods as " . . . *methods that are used to detect the presence of a substance or class of substances at the level of interest. These methods have the capability for a high sample throughput and are used to sift large numbers of samples for potential non-compliant results. They are specifically designed to avoid false compliant results*". Data from screening assays are normally qualitative or semi-quantitative.

In contrast, according to EU legislation[7] a "confirmatory method" means "*methods that provide full or complementary information enabling the substance to be unequivocally identified and if necessary quantified at the level of interest*". Current confirmatory assays for hormones are predominately based on either Liquid-Chromatography-Mass Spectrometry/Mass Spectrometry (LC-MS/MS) and/or a variant of Gas Chromatography-Mass Spectrometry (GC-MS/MS).

Method development for the reliable detection of residues of hormones has been going on for many decades. Next to screening methods, instrumental confirmatory techniques are a necessary tool for control laboratories. Furthermore, with the introduction of fast chromatographic techniques such as Ultra Performance Liquid Chromatography (UPLC), many laboratories now choose to employ the same mass spectrometric technique for both confirmation and screening. All of these methods are capable of providing quantitative data, and meet the demanded criteria for analyte identification.[7]

Typically more than 20 related hormones are included in a single mass spectrometric method.

More recently, other techniques such as Time-of-Flight Mass Spectrometry (ToF-MS) and Gas Chromatography-Combustion-Isotope Ratio Mass Spectrometry (GC-C-IRMS) have been employed in hormone analyses. The move from targeted methods (based on techniques such as LC-MS/MS) to accurate mass full-scan MS techniques, such as ToF-MS, provides the exciting opportunity to detect many more analytes (>100) in a single analytical run. At present the ToF-MS technique tends to be less sensitive than LC-MS/MS, and is therefore more suited to either matrices that contain higher concentrations of residues, *e.g.* urine, or those samples which have undergone more extensive clean-up and concentration of the final extract. Whilst the GC-C-IRMS technique is well established in the field of sports doping[8] it is still a relatively new addition to the field of animal testing.[9] This technique typically utilises the $^{13}C/^{12}C$ ratio between an individual hormone and a suitable Endogenous Reference Compound (ERC), *e.g.* dehydroepiandrosterone (DHEA), to help differentiate between an administered natural hormone and the natural background concentration within an animal. More details on this technique can be found in Chapter 5.

This chapter describes the evolution of analytical methods for the detection and identification of residues of hormones in biological matrices. Most early methods used in Europe were focused on the testing of hormones in urine due to the relatively high concentrations of hormones present in this type of sample and the ease of "on-farm" sampling. More recently, other non-invasive matrices have also been investigated, *e.g.* hair, as reservoirs of hormone esters, in an attempt to identity hormone (ab)use over a longer time period and to detect the hormones in the chemical form they are administered in order to provide proof of treatment in the case of endogenous hormones. This chapter also describes the "state-of-the-art" for the analysis of muscle tissue, since this is one of the greatest challenges for residue chemists due to the very low concentrations that can be expected in this matrix. In fact many EU Member States do not give a high priority to this matrix for their National Control Programmes because of the relatively high costs involved and the availability of less costly alternatives when the animal is still alive or at the slaughterhouse. However, for import control and ultimate consumer safety the analysis of meat is necessary. The performance of confirmatory methods for hormone testing in Proficiency Testing Schemes (PTs) is also reviewed.

4.2 Biochemical Screening Methods

4.2.1 Immunological Procedures

There are many commercial ELISA kits available for detection of hormone residues in animal tissues and urine. Table 4.1 presents some information on

Table 4.1 Examples of commercially available ELISA kits for detecting hormone residues.

	Manufacturer claimed detection limit(s) (μg kg^{-1} or μg l^{-1})							
Analyte	*Bile*	*Faeces*	*Kidney fat*	*Milk*	*Muscle*	*Perineal fat*	*Plasma*	*Urine*
17β-Oestradiol							R(~0.02)	
Boldenone								T(1)
Diethylstilboestrol	R(~2)	R(~0.1)			R(~0.1), T(0.1)			A(0.2), E (0.5), R(~0.2), T(0.2)
Ethinylestradiol							R(~0.02)	R(~0.2)
Medroxyprogesterone acetate			E(0.5)					
Melengestrol acetate					R(0.075)	R(0.3)		
Methyl testosterone								R(~0.2)
Nortestosterone					T(0.0125)			A(0.5), E(0.5), R(3), T(0.125)
Progesterone				A(0.1), E(0.1)				
Testosterone							R(0.02)	
Trenbolone	R(~1)	R(~0.025)			R(~0.2)			A(0.5) E(0.5), R(~0.4), T (0.125)
Zeranol					T(0.05)			A(0.1), E(0.1), R(~1.5), T(0.25)

A=Abraxis (http://www.abraxiskits.com/product_veterinary.htm), E=Euro-Diagnostica BV (http://www.elisa-tek.com/eurodiagnostica%20product%20list.htm), R=R-Biopharm Rhone Ltd (http://www.r-biopharm.com/), T=Tecna (http://www.tecnalab.it/itenglish/home). Information collated from company websites as of March 2009.

currently available kits, including the applicable matrices and claimed detection limits.

In general all of these commercial kits are capable of meeting the recommended concentrations proposed by the EU Community Reference Laboratories (Table 4.2).[10] It should be noted, however, that the available ranges of kits do not provide coverage of all of the current analytes of interest. Nevertheless, while many regulatory laboratories prefer to use multi-residue methods based on mass spectrometry, ELISA methods are routinely employed by a number of EU Member States as a first action in their National Residue Control Plans (NRCPs). Suspect positive (or potential "non-compliant"[7]) samples from these ELISA screening assays are then confirmed by either GC-MS(MS) or LC-MS/MS.

Recent examples of the use of ELISA in monitoring programmes include the UK where individual kits are used to monitor imported beef for residue of trenbolone (TREN) and zeranol (ZER).[11] Other ELISA kits are also used to monitor for progesterone and testosterone in samples of UK production as part of the NRCP.[12] ELISA-based methods are also used in many other EU Member States, including the Netherlands, Germany and Spain, as part of a suite of methods to detect stilbenes, steroids and resorcylic acid lactones (RALs), *e.g.* ZER and associated compounds.

While there was early interest from both Canadian and USA regulatory laboratories in the development of RadioImmunoAssay (RIA) for ZER, the primary focus was on the development of GC and GC-MS methods, the technologies already used for testing for diethylstilboestrol (DES). Since then the USDA/FSIS has included the use of a commercially available ELISA test kit for ZER residues in liver or muscle in its NRCP.[13] Homogenised tissue is incubated with glucuronidase to release bound residues, then further extracted and cleaned up by partition with organic solvents. The final extract for ELISA assay is in an aqueous buffer. ZER residues are detected at concentrations of $0.5\,\mu g\,kg^{-1}$ or higher.

The USDA/FSIS Chemistry Laboratory Guidebook currently includes an ELISA method for screening for melengestrol acetate (MGA) residues[14] and an LC-MS/MS method for residue confirmation.[15] The ELISA, which uses proprietary C18 SPE (Solid Phase Extraction) cartridges for clean up of fat samples prior to measurement using a proprietary test (Ridascreen® MGA ELISA), is validated for detection of residues at concentrations $\geq 10\,\mu g\,kg^{-1}$. Lipids are removed from extracts prior to SPE clean up by chilling and centrifugation. The bound MGA conjugate is blue in colour and is detected at 450 nm. The ELISA method Standard Operating Procedure (SOP) includes a warning under preparation of standards that MGA may degrade when exposed to ultraviolet light and that standards should be prepared and stored in amber containers, with recommended storage at typical refrigeration temperatures (2–8 °C).

The use of other immunological-based techniques in hormone testing is extremely limited. For example, whilst optical biosensor methods based on Surface Plasmon Resonance (SPR) have been developed for antibiotics and

Table 4.2 Recommended concentrations proposed by the EU Community Reference Laboratories

		Recommended concentration ($\mu g\,kg^{-1}$ or ($\mu g\,l^{-1}$)				
Analyte	*Marker residue*	*Kidney fat*	*Liver*	*Muscle*	*Serum*	*Urine*
Diethylstilboestrol	Diethylstilboestrol		2	1		1
Dienoestrol	Dienoestrol		2	1		2
Hexoestrol	Hexoestrol		2	1		2
Boldenone	17β-boldenone conjugate		2	1		1
17β-19-nortestosterone (i)	17α-19-nortestosterone (ii)		2	1		1
Ethinylestradiol	Ethinylestradiol		2	1		1
17β-Oestradiol	17β-Oestradiol			1	0.1	
17β-Testosterone	17β-Testosterone				Male < 6 months 10, Male 6–18 months 30, Female < 18 months 0.5	
Methyltestosterone	Methyltestosterone		2	1		2
17β-trenbolone	17α-trenbolone		2	1		2
Stanozolol	16β-hydroxystanozolol		2	1		2
Dexamethasone	Dexamethasone		MRL(iii)	MRL(iii)		2
Megestrol	Megestrol acetate	5		1		
Melengestrol	Melengestrol acetate	5		1		
Chlormadinone	Chlormadinone acetate	5		0.5		
Medroxy progesterone	Medroxy progesterone acetate	1 (MRPL)		1		
Zeranol	Taleranol		2	1		2

*CCβ for screening methods or CCá for confirmatory methods should be lower than the CRL recommended concentrations. Blank spaces in table=no recommended concentration.

(i) 17β-19-Nortestosterone occurs naturally in non-castrated pigs and horses. (ii) 17α-19-Nortestosterone occurs naturally in pregnant cows and newborn calves. (iii) When there has been an authorised treatment.

β-agonists,[16] there are no commercially available SPR kits for the detection of hormone residues. Nevertheless, this is an active area of research and an EU Framework 6 funded project, "BIOCOP" (New Technologies to Screen Multiple Chemical Contaminants in Foods), is currently developing a new SPR biosensor assay for detecting hormone growth promoters *via* protein biomarkers in blood samples.[17] In this project it is anticipated that biomarkers to hormone treatments will provide a novel way of differentiating between potentially illegally treated and untreated animals. It is also believed that newly developed illegal designer steroids might also be traceable using biomarker screening.

4.2.2 Receptor-based Assays

4.2.2.1 Oestrogens (REA)

Whilst receptor-based assays have been used for some time to detect dioxins and polychlorinated biphenyls (DR-CALUX®) the potential of this type of method for testing of hormones in animal and feed samples is a relatively new development. In the case of oestrogenic compounds, Bovee[18] reported the validation of a rapid yeast oestrogen assay (REA) for the screening of oestrogenic activity in calf urine in 2005. This method, which was based on the expression of a yeast-enhanced green fluorescent protein (yEGFP), was capable of detecting low concentrations of a wide range of oestrogens such as 17β-oestradiol ($1\,\mu g\,l^{-1}$), diethylstilboestrol ($1\,\mu g\,l^{-1}$), 17α-ethynylestradiol ($1\,\mu g\,l^{-1}$), α-zearalanol ($50\,\mu g\,l^{-1}$) and mestranol ($10\,\mu g\,l^{-1}$) in calf urine. High concentrations of the androgen testosterone and the gestagen progesterone were shown not to give a response in this bioassay. This approach, however, was not applicable to urine samples from adult animals as the REA is hindered by the presence of natural steroids such as 17α-oestradiol and estrone; at high concentrations of 17α-oestradiol and/or estrone, all urines would be screened suspect for oestrogens.

The REA approach method has also been used to further investigate a number of "suspect" samples with unusual oestrogen activity. In this case urine extracts were fractionated using a Liquid Chromatography (LC) system into a 96-well plate. Those wells giving a response on the biosensor were then correlated to LC retention time. The eluent from the LC column at that retention time was also investigated using hybrid quadrupole/orthogonal Time-of-Flight Mass Spectrometry (QToF-MS/MS) which provided exact mass measurements. Data from this combined bioassay-MS approach were used successfully to identify residues of both known and unknown oestrogens in calf urine.[19] In 2006 a further two papers on the use of the oestrogen receptor assay were reported.[20,21] The first, which described the extension of the technique to animal feed, was capable of detecting 17β-oestradiol, 17α-ethynylestradiol and diethylstilboestrol, at 5, 5 and $10\,\mu g\,kg^{-1}$, respectively. Zearalanone and equol could only be detected at much higher concentrations of 1.25 and $200\,mg\,kg^{-1}$. Once again the assay was not affected by 17β-testosterone or progesterone.

Whilst the method was generally robust, it was, however, noted that a small number ($<1\%$) of feed samples were cytotoxic to the assay. The second paper[21] provided data on a comparison between the data for calf urine obtained by REA and that obtained by GC-MS/MS. Over a two-year period negative and positive calf control urine samples were analysed. When compared to the established GC-MS/MS method the REA showed a relatively low rate of potential false-suspect results ($\sim 6\%$). Whilst the REA can be regarded as a fast and reliable assay, one of the potential drawbacks preventing the wider use of this technique is the need to handle the recombinant yeast in special laboratory facilities.

4.2.2.2 Androgens (RAA)

To date the use of receptor assays to detect androgenic compounds in food and animal feed has only received limited attention and, as yet, has not been incorporated into the routine screening tests employed by laboratories. In fact the most recent developments in this area were reported in 2007/08 by Bovee *et al.*[22,23] Using a similar approach to the REA, this work has focused on the construction of a yeast-based androgen bioassay that expresses the human androgen receptor (hAR) and yEGFP as a measurable reported protein in response to androgenic compounds. As with all bioassays the "cross reactivity" (expressed as relative androgenic potencies, RAPs – defined in this study as the ratio between the EC_{50} of 17β-testosterone and the EC_{50} of the test compound) varied according to chemical structure and the ability of that structure to bind to the receptor. For example, when using solutions of chemical standards, 5α-dihydrotestosterone (RAP=2.3), 17β-testosterone (RAP=1.0) and 17β-boldenone (RAP=0.15) were all found to be potent androgens producing a dose-related increase of the green fluorescent protein. In contrast 17α-testosterone did not elicit a response, indicating that the 17-β-OH function is very important with regards to androgenic activity. This assay was also shown to detect the female hormone 17β-oestradiol (RAP=0.0084), but only at concentrations *c.* 500 times that of 17β-testosterone. This procedure was reported to be fast, sensitive and a very specific assay to compounds that have an antiandrogenic mode of action. Whilst this technique offered the potential to detect new or "designer" steroids it was not capable of detecting prohormones (or steroid precursors) with an androgenic mode of action, *e.g.* dehydroepiandrosterone (DHEA), and this limitation was addressed by Rijk in 2008.[24] In this study prohormones were activated immediately prior to the RAA. This typically involved either chemical standards or purified sample extracts being incubated with bovine liver S9 at 37 °C for 6 hours. The developed system was shown to mimic the hydroxysteroid dehydrogenase and cytochrome P450 mediated *in vivo* metabolic transitions and provided a mechanism for both bioactivity and steroid identification without the need for animal studies. Whilst this proved that DHEA and other prohormones can be converted into more potent androgens to aid detection, it could be concluded that this technique is not yet ready for use in routine National Residue Control Plans and is only really suited for detecting prohormones in feed supplements and other types of steroid preparations.

4.3 Chromatographic Methods

4.3.1 Stilbenes

Testing for compounds with hormonal-based growth promoting activities began in North America with the use and subsequent ban of DES in food animal production in the 1970s. The primary analytical technique used in both USA and Canadian regulatory laboratories at that time was gas chromatography with electron capture detection for quantitative determination, with confirmation by GC-MS. The early work on DES confirmation has had a profound effect on the manner in which confirmation of an analytical result is conducted today. In 1977, the US Food & Drug Administration established a requirement that assays used for the detection of carcinogenic residues must have demonstrated specificity.[25] In response to this requirement, a study was conducted to establish the criteria required to demonstrate specificity for the analysis of DES residues using GC-MS with ionisation by electron impact (EI).[26] Early methods typically involved extraction, liquid partitioning and derivatisation for electron capture GC or electron impact GC-MS determination.[27–29] With the evolution of technologies in the late 1970s and the introduction of SPE and high-performance liquid chromatography (HPLC, or now more commonly abbreviated as LC) into routine use in residue control laboratories, methods typically consisted of steps which included enzymatic deconjugation using *Helix pomatia*, SPE using C18-cartridges, isolation of DES from the extract with LC and preparation of trimethylsilyl (TMS) derivatives for GC-MS detection and confirmation. For initial screening the molecular ion of DES-TMS at m/z 412 was monitored, together with the isotope ion at m/z 413. This ratio had to fall within a predetermined range. For confirmation a separate high resolution (hr) measurement was made in which the measured mass had to be within ± 0.0012 u of the theoretical value of 412.2254.

In the Sphon method[26] the spectrum of DES was matched against spectra in a library which at the time contained GC-MS electron impact spectra for approximately 30,000 compounds. It was demonstrated that monitoring the three major peaks in the DES spectrum for both presence and relative intensity provided a unique match to the DES spectrum in the library. This study was repeated at an ASMS workshop in 1996, by which time the available libraries contained about 270,000 EI spectra.[30] Again, monitoring the three characteristic major ions from the DES spectrum produced no other matches when tested on the basis of ion presence and ratio of selected ion intensity to base peak intensity. From that early work has developed the current regulatory practices as recommended by the American Society for Mass Spectrometry[31] and as contained in European Union requirements.[7,32]

A fully validated confirmatory analytical method for DES was developed and published in Europe in 1986.[33] This was based on gas-chromatography mass spectrometry, using isotope enriched (deuterium) DES (DES-d_6) as internal standard (GC-IDMS). The limit of detection at which confirmation of the identity was possible was in the range of 1–2 $\mu g\,l^{-1}$. This procedure was validated against a radioimmunoassay during which a good correlation was found in the resulting quantitative values. No further validation data of the method were reported.

Within Europe the early GC-MS methods[26,33] have not been widely used. The main reason was the lack of applicability of the hrMS-systems in the late 1980s. Robustness of the equipment was poor and running this type of instrument was a highly specialised discipline. It took until the early 1990s, when relatively simple low resolution (lr) machines with unit mass resolution became available, for MS to become a widely accepted technique, routinely applicable for residue analyses. The development of new methods for residues of DES and the other stilbenes, hexoestrol (HEX), dienoestrol (DE) and benzoestrol (BEZ), has been an ongoing activity since that time, resulting in several sensitive methods. These methods frequently are based on GC-lrMS equipment such as GC-MSD (Mass Selective Detection).[34,35]

Recently the EU Community Reference Laboratory, RIVM, finalised a new method for stilbenes in biological materials.[36] This time the method was based on GC-MS/MS for detection and confirmation of the identity and was deemed suitable for all stilbenes currently of relevance for residue control. This new method was developed and validated for both urine and muscle tissue. In brief, meat/urine is spiked with internal standards DES-d_6, HEX-d_4, DE-d_2. Homogenised muscle is extracted with water using ultrasound and, following addition of methanol, defatted using *n*-heptane. For urine samples, acetate buffer (pH=5.2) and β-glucuronidase/sulfatase are added prior to incubation overnight at 37 °C or, alternatively, 2 hours at 50 °C. After cooling, analytes are extracted using tertiary butyl-methyl-ether (TBME). Stilbenes of interest are isolated using an analytical LC column with a C18 phase. The dried fractions from the LC are derivatised using MSTFA^{++} [a mixture of N-methyl-N trimethylsilyltrifluoroacetamide (MSTFA): ammonium iodide: dithioerythreitol] prior to measurement. Screening analyses are performed *via* GC-MS. When confirmation of the identity is necessary, these measurements are performed by GC-MS/MS under the conditions described below (Table 4.3).

This GC-MS/MS method was validated according to the latest EU method validation guidelines[7] resulting in the following performance characteristics: Detection Capability (CC_β) for DE (0.14 $\mu g\,kg^{-1}$), HEX (0.09 $\mu g\,kg^{-1}$), BEZ (0.14 $\mu g\,kg^{-1}$) and DES (0.10 $\mu g\,kg^{-1}$). The validation of the method's ability to confirm analyte identity was achieved by analysing samples spiked at the EU Recommended Concentration for testing.[10] Confirmation using the four Identification Point (IP) criteria[7] proved to be possible at concentrations of $\leq$0.5 $\mu g\,kg^{-1}$.

Thus, for stilbenes, it appears that the progress made over the period of roughly two decades has been mainly with respect to the robustness and

Table 4.3 GC-MS-MS measured transitions

Analyte	*MRM I*	*MRM II*
Dienoestrol	410 > 381	410 > 395[a]
Hexoestrol	207 > 179	207 > 191[a]
Hexoestrol-D4	209 > 193[b]	
t-Diethylstilboestrol	412 > 217[a]	412 > 383
t-Diethylstilboestrol-D6	418 > 220[b]	
Benzestrol	207 > 179	207 > 191[a]

[a]Trace used for quantification
[b]internal standard.

applicability of the analytical methods. The analytical power of the old and new methods [in terms of Limit of Detection (LOD) or Detection Capability (CCß)] on the other hand is not that different.

4.3.2 Melengestrol Acetate (MGA)

Another hormonal compound, melengestrol acetate (MGA), registered for use as a feed additive in Canada and the United States of America, has also been included in the routine residue testing since the early 1980s. The primary analytical method at that time used Gas-Chromatography (GC) on a packed column with Electron Capture Detection (ECD). GC-MS was employed as the confirmatory technique. The GC-ECD method for MGA was as provided by the registrant of the drug, but the method was also published following a collaborative study in the open scientific literature[37] and subsequently in the AOAC Official Methods of Analysis as Method 976.36, applicable to bovine liver, kidney, muscle and fat, with a limit of quantification of 10 $\mu g\,kg^{-1}$.[38] The usual target tissue for regulatory analysis was fat, where residues of MGA are most prevalent.[39] An updated version of the method using capillary GC continues to be listed as the approved method for use in the USDA/FSIS residue testing programme.[40] The current version of the method is validated for the quantitative determination of MGA residues in bovine fat at concentrations from 10–30 $\mu g\,kg^{-1}$, with recoveries in the range of 70–115% and repeatability of ~20%.

By the mid 1990s, changing technology and requirements to enhance capabilities led to the development and implementation of an LC-Ultraviolet (UV) method for MGA and the related progestagens, megestrol acetate and chlormadinone acetate (CMA).[41] This method has been validated for analysis of residues of the three progestagens in bovine fat found in edible tissue at concentrations between 10 and 1000 $\mu g\,kg^{-1}$. Fat is first rendered, then extracted with acetonitrile and the extract is dried after washing with hexane. After removal of lipids by saponification and precipitation, the progestagens are extracted from the remaining basic solution with hexane. The progestagens are recovered from a cyanopropyl SPE column and determined by LC on a C18 column with UV detection at 291 nm. Reported recoveries from fortified samples are 84–116%; with a detection limit of 3 μg kg^{-1}. Positive findings are confirmed by re-analysis, using LC/MS/MS.[42]

USDA scientists and collaborators have developed two LC methods for MGA which have potential regulatory use, but neither is currently contained in the USDA/FSIS Chemistry Laboratory Guidebook.[43] The first of these methods was developed by researchers at Cornell University, in collaboration with the USDA, and used liquid-liquid extraction of tissue homogenates followed by analysis using coupled phenyl and silica analytical columns (normal-phase) for LC with UV detection at 287 nm.[44] Reported recovery was 86% with repeatability of ~10% at the 10 μg kg^{-1} concentration in bovine liver extracts. Subsequently, researchers with the USDA Agricultural Research Service (ARS) investigated Supercritical Fluid Extraction (SFE) followed by SPE for the analysis of MGA residues in bovine fat.[45] Final extracts were suitable for analysis by either LC with UV detection or GC-MS. Reported recoveries from bovine fat were 99% with a Coefficient of Variation (%CV) of 4%. The MGA confirmatory method based on LC-MS/MS uses the same extracts from the GC method which, after drying, are taken up in LC mobile phase (methanol: water, 55:45, plus 0.1% formic acid) for MS/MS analysis. The method monitors for the precursor ions 397 and 337 and product ions 319 and 279, using the ratio 319/279 for confirmation (to agree within 20% of the ratio obtained from injections of a standard).

In Europe the status of MGA is not different from that of the other gestagens like medroxyprogesterone acetate (MPA), CMA and megestrol acetate. For confirmation, initially GC-MS was used but gradually there has been a shift to LC-MS/MS methods. Several approaches have been used for the analyses of both fat and muscle tissue. One approach was based on the SFE method developed at the USDA ARS[45] in combination with LC-MS/MS.[46] Alternatively, another method was developed in which an organic solvent extract was only purified using SPE after which the extract was directly analysed with LC-MS/MS.[47]

4.3.3 Trenbolone

The first experience in Canadian regulatory laboratories with non-approved use of hormones for growth promotion occurred in late 1985 and early 1986. Prior to the approved use of trenbolone acetate in feedlot beef cattle in Canada, Canadian Customs officers identified illegal importation of TREN implants into Canada, leading to increased surveillance by meat inspection staff at federally inspected plants for evidence of use. Subsequently, implants found in the brisket area of a veal calf at slaughter were identified as containing trenbolone acetate, a product then approved for use in veal production in some European countries. As actual implants were found by inspection staff and submitted for laboratory analysis, residue methodology was not required. The contents of the implant capsules were dissolved and analysed by GC-MS, using standard derivatisation procedures from published methods of the time, but there was no formal publication of the method, the use of which was considered as "non-routine". Subsequent testing of implants found in veal calves sampled during the surveillance programme did not reveal the use of any hormones other than trenbolone. This method continued to be used for several years whenever inspectors found implants which were not readily identifiable or were not implanted at the approved site. This first experience of Canadian Regulatory Laboratories with the use of non-approved growth-promoting hormones and related substances led to further analytical methods being developed as new information on such misuse became available.

The methodology adopted by USA and Canadian Regulatory Authorities for trenbolone acetate residues, following the approval of use as an implant in feedlot cattle in both countries, was developed by at Cornell University with support from USDA/FSIS.[48] The original method, which detects both α-TREN and β-TREN, uses 19-nortestosterone as an internal standard. Following a three-phase liquid extraction, extracts are cleaned up by SPE and analysed by LC with UV detection at 350 nm. Conditions were provided in the reported method for LC-MS/MS confirmation. A second method reported by the Cornell group used GC-MS, with co-injection of sample extract and derivatising agent as reported in the ZER method.[49] Although the method development was also sponsored by USDA/FSIS, this method was not adopted for routine regulatory use. The LC-UV method was routinely used until concerns were raised in an international audit concerning the use of 19-nortestosterone as an internal standard, as evidence of misuse of this compound as an illegal growth promotant had been reported in Europe.[50–54] In addition, there was a subsequent finding that 19-nortestosterone was naturally present in certain species and classes of animals.[55–61] The method was subsequently modified to remove the use of 19-nortestosterone as an internal standard for use in Canadian regulatory control testing.[62]

Since the EU's complete ban on the use of TREN in 1988 there have been relatively few reports of its abuse. Nevertheless, TREN analyses have always triggered the development of new approaches. As in the USA, the UV absorbance characteristics have been the basis for analytical methods with the EU. However, in the early days of GC-MS, methods were developed on the basis of Immuno Affinity Chromatography (IAC) as a sample clean-up procedure prior to detection with GC-MS.[63] In the following years several methods were developed, with a gradual change from GC to LC.[64–66] Very recently a method was published using accurate mass measurement with LC-ToF-MS, using a similar IAC sample purification approach to that used 20 years earlier for one of the first GC-MS methods.[67]

4.3.4 Multi-residue Hormone Methods using GC-MS and/or LC-MS/MS

4.3.4.1 Background

Though well established and generally accepted, the EU Identification Point (IP) approach[7] places heavy demands on the analytical methods used for analyte confirmation in hormone analyses. The necessity of collecting four IP for confirmation of the identity of a banned substance, when using *e.g.* GC-lrMS, translated into the detection of four MS signals (four diagnostic ions), all with an S/N ratio which allows quantification. In practice, most of the published methods had no problems with the detection of two to three ions, but four frequently becomes difficult. Moreover, the relative response ratios for these diagnostics ions have to fit those obtained for the reference compound. Stolker *et al.*[68] studied the power of a well-established analytical procedure for five hormones (methylboldenone, methyltestosterone, ethynylestradiol, boldenone and nortestosterone) in four different muscle tissue matrices (cattle, turkey, fish and pork). Their conclusion was that under routine conditions (single-shot analyses for a specific sample) at a concentration of 1 $\mu g\,kg^{-1}$, confirmation is only possible in approximately 50% of cases. The "drive" for residue chemists all over the world is therefore to continue the development of new and improved methods based on three different objectives:

- The development of routinely applicable methods, suitable for both screening and confirmation of analyte identity;
- The move from targeted methods to non-targeted (generic) methods, based not only on biochemical and biological principles, but also by the use of instrumental techniques;
- The inclusion of more analytes in a single procedure.

In order to include more and more different analytes in a single assay the selectivity of sample clean-up procedures requires optimisation for compatibility with the final detection technique *e.g.* LC-MS-MS. If the extraction/purification technique is too specific, compounds of interest may not be injected into the MS system. Conversely, if the purification process is insufficient, matrix interferences and ion suppression in the MS may also result in unreliable data.

One purification technique used in the early 1990s was that of Multi Immunoaffinity Chromatography (MIAC). In this approach, several antibodies raised against specific steroid hormones were combined into a single softgel column, allowing the highly specific isolation of different hormones from single crude extract.[63] Despite its elegance, this approach suffered from two drawbacks. First the availability of constant supplies of high-quality antibodies; second the fact that these antibodies frequently are too specific in those cases where slightly different molecules have to be introduced into the analytical procedure. Most antibodies were raised against 17β-nortestosterone (nandrolone). However, when the primary metabolite in bovine urine was identified as 17α-nortestosterone, the analytical procedure used for muscle tissue was not suitable due to the low cross reactivity of the antibody, raised against 17β-nortestosterone, for 17α-nortestosterone. More recently, the improved availability of high-quality SPE materials and the technological advances in MS have facilitated the further development of multi-residue methods for hormones.

4.3.4.2 North America

During the 1980s, with the approval in a number of countries (including the United States of America and Canada) of the use of hormonal growth promoters containing synthetic versions of the natural hormones oestradiol, progesterone and testosterone, as well as the synthetic hormonal compounds ZER and trenbolone acetate, the focus of methods development shifted to these latter two compounds. Evaluations of the "natural" hormones oestradiol, progesterone and testosterone at both national and international levels had led to the conclusion that their use as directed did not lead to elevated concentrations of these hormones in meat from animals implanted with the pellets.[69] As a result, regulatory authorities in Canada and the USA did not perceive a requirement or a benefit to be gained from expending resources on the development and routine implementation of analytical methods for these three compounds. The focus of methods development was therefore on the two synthetic hormones administered *via* implants, trenbolone acetate and ZER, on the synthetic hormonal feed additive MGA and on the banned compound DES.

Although a method was published by a US Food & Drug Administration laboratory for analysis of ZER by LC with electrochemical detection, with confirmation by GC-MS,[70] a method developed for the US Department of Agriculture's Food Safety & Inspection Service laboratories at Cornell University's Equine Drug Testing and Toxicology Laboratory became the accepted regulatory method for ZER and DES in both the United States of America and Canada in 1986.[71] The method continues to be used in Food Safety and Inspection Service Laboratories in the USA for analysis of ZER and DES residues[72] and in Canadian regulatory testing laboratories. As originally published, the method was intended for use in screening, determination and confirmation for oestrogenic compounds in bovine liver, kidney or muscle. Compounds included in the original method were DES, DE, HEX, ZER, taleranol, zearalanone, zearalenol, oestradiol and oestriol. To meet the criteria which required that a regulatory method accepted for use in the USA for analysis for veterinary drug residues should successfully complete a multi-laboratory trial involving a minimum of three laboratories, a method trial was conducted involving laboratories in the USA and Canada, demonstrating the successful application of the method to the two analytes of primary concern, ZER and DES. It should be noted that this does not mean that the trial was not successful for the other compounds, but simply that the method was tested only for these two analytes in the multi-laboratory trial. At the time this work was done, DES use had been banned in food animals in Canada and the USA, with the parent compound identified as the marker residue. ZER had an approved use, but the Maximum Residue Limits (MRLs) were expressed in terms of zeranol as the marker residue. Taleranol and associated residues were not included in the marker residue definition and therefore were not considered as key analytes for validation in the multi-laboratory trial.

The method included digestion with glucuronidase to release bound residues, followed by a three-phase extraction, with the middle phase containing the extracted hormones. Clean up was conducted on SPE cartridges using a modified centrifuge to achieve rapid separation and the final extract was injected into the GC-MS in combination with the derivatising reagent. Zeranol has multiple active sites, which can react with the derivatising agent, but reaction in the vapour phase in the injection port provided a stable and reproducible reaction. Attempts to derivatise ZER using the more conventional approach of reaction in the liquid phase prior to injection can prove frustrating due to lack of reproducibility both within and between runs even though the same reaction conditions are applied to all replicates.

The method has also received recognition as suitable for supporting the Codex Alimentarius MRLs established for ZER residues in bovine muscle and liver.[73] Additional within-laboratory validation work was conducted by the Centre for Veterinary Drug Residues of the Canadian Food Inspection Agency, resulting in the extension of the validated method as used in Canada to include the two other stilbenes, DE and HEX, and ZER-associated compounds such as taleranol and zearalenone, in part to address method requirements for products exported to EU Member States. The regulatory status of ZER in the EU and in North America is different (banned as a hormone in the EU, approved for specified use in Canada and the USA), leading to different analytical method requirements to address these respective regulatory environments. Any ZER-associated residue finding becomes a reason for investigation under EU legislation, while only an excess of the marker residue in beef tissues or detection of residues in meat other than beef are grounds for regulatory action in Canada and the USA.

To address also EU requirements for live animal testing, the method for ZER and stilbenes was adapted by Canadian regulatory scientists to test urine from beef cattle raised under a "hormone-free" programme to qualify for EU market access.[74]. The revised method uses commercially available immunoaffinity columns to achieve separation of the analytes, following which the extracts are analysed by GC-MS using the same procedures developed for analysis of tissue extracts. Three ions are monitored for each of the target analytes, with a requirement that ion ratios must match the equivalent ratios for pure standards. Zearalanane and DES-d_8 are used as internal standards. The method was validated to permit detection of the target analytes at a minimum concentration of $2\ \mu g\ l^{-1}$. Ions monitored and Detection Capability ($CC\beta$)[7] for the four target analytes included in the urine method are given in Table 4.4.

Table 4.4 Ions monitored, analytical recovery and detection capability CCβ for zeranol and stilbenes (as TMS derivatives) in Canadian Food Inspection Agency method applicable to urine.

Analyte	*Ions Monitored (m/z)*	*Recovery (%) at $2\ \mu g\ l^{-1}$*	*CCβ ($\mu g\ l^{-1}$)*
Zeranol	433, 523, 538	96	0.28
Diethylstilboestrol	412, 383, 397	98	0.24
Dienoestrol	410, 395, 381	81	0.15
Hexoestrol	207, 191, 414	100	0.84

The USDA/FSIS *Chemistry Laboratory Guidebook* does not currently include any test methods for the endogenous hormones oestradiol, progesterone and testosterone and testing for these compounds is not included in the "Blue Book" plan for 2007, the most recent year available on the USDA/FSIS website.[75]

Methods currently used in Canada's national residue programme include a GC-MS method for 17*β*-oestradiol in urine,[76] an LC-MS/MS method for progesterone, testosterone and epi-testosterone in muscle and liver tissues[77] and an LC-MS/MS method for hormone esters in suspected injection sites.[78] Although Canada had taken the same approach to endogenous hormones as the USA in not including these compounds in the NRCP until after 2000, commitments made to the EU to support a potential export of "hormone-free" beef required development and implementation of tests for these compounds. These tests also were required to support the EU regulatory approach, which requires both testing of live animals while on the farm, using urine, as well as testing of meat collected at slaughter. Initial methods development therefore was split between investigation of a method for endogenous hormones in urine by GC-MS, using 17*β*-oestradiol as the initial target analyte, while the focus for a tissue method was on testosterone and progesterone, to thereby provide methods for all three endogenous hormones in the testing programme, particularly for cattle registered in the "hormone-free" programme.

Development of the oestradiol method for urine was based on a previously developed method for 17*β*-oestradiol and 17*β*-TREN residues in bovine serum.[79] The method developed for serum, as a potential alternative for urine testing, included initial extraction with 1-chlorobutane, clean-up on a silica gel SPE cartridge and derivatisation with pentafluorobenzoyl chloride. GC-MS analysis was conducted using a DB-5MS column, with fragment ions generated in the chemical ionisation mode using methane as the buffer gas. The ions monitored are m/z 464 for 17*β*-TREN and m/z 664 for 17*β*-oestradiol, with deuterated internal standards used for both compounds. The linear range was 5–500 $ng\,l^{-1}$ for 17*β*-TREN and from 25–2500 $ng\,l^{-1}$ for 17*β*-oestradiol. Calculated values for the Decision Limit ($CC\alpha$)[7] and $CC\beta$ are given in Table 4.5.

The method, as modified for application to urine, required additional steps due to the formation of an emulsion when 1-chlorobutane was added directly to urine.[76] In the urine method buffer is added to the urine test portion, followed by addition of glucuronidase and overnight incubation to release any bound residues. The samples were then cleaned up using an OASIS™ HLB SPE cartridge, which contains a hydrophilic-lipophilic balanced phase packing

Table 4.5 Decision limits and detection capabilities for determination of 17β-oestradiol and 17β-trenbolone in bovine serum by GC/MS

Analyte	*CCα (ng l^{-1})*	*CCβ (ng l^{-1})*	*Coefficient of variation across linear range (%)*
17β-oestradiol	10.3	49.3	< 10%
17β-trenbolone	17.3	82.7	< 12%

material. Oestradiol was eluted from the SPE cartridge with 10% methanol in MTBE and the volume was reduced to 0.5 ml, which was then extracted with three 2-ml portions of 1-chlorobutane, which were combined and taken to dryness. The dried extract was then derivatised with pentafluorobenzoyl chloride and analysed by GC-MS, using d_4-17β-oestradiol as internal standard. A linear response was obtained between concentrations of 100 and 1000 $ng\,l^{-1}$, with recoveries ranging from 80–130% and calculated values of 170 $ng\,l^{-1}$ for CCα and 287 $ng\,l^{-1}$ for CCβ. Although 17β-oestradiol is metabolised to 17α-oestradiol and excreted in urine, administration of 17β-oestradiol in implants has been reported to result in increases in concentration of the 17β-isomer in urine.[80] Monitoring for 17β-oestradiol in urine is therefore an indicator of a non-endogenous source.

The method developed for endogenous hormones in tissue was initially validated and implemented for residues of progesterone, testosterone and epi-testosterone in bovine muscle and liver.[77] After initial denaturation of endogenous enzymes with methanol, dried samples are buffered and digested overnight with glucuronidase, then extracted with MTBE. The dried extract is dissolved in acetonitrile and cleaned up with a series of partitioning steps followed by separation of the analytes on a silica gel SPE cartridge and LC-MS/MS analysis. For progesterone, the transitions monitored are (m/z) 315 to 109 and 315 to 97, while transitions for testosterone and epi-testosterone are 289 to 109 and 289 to 97. Deuterated testosterone is used as internal standard. The detection limit and CCα for these compounds were calculated to be 0.5 $\mu g\,kg^{-1}$, while CCβ was calculated to be 0.8 $\mu g\,kg^{-1}$. The upper limit for these endogenous hormones is reported to be about 0.4 $\mu g\,kg^{-1}$,[81] so the method was considered suitable for the detection of elevated concentrations which might result from external sources and was validated to meet confirmation requirements of Commission Decision 2002/657/EC.[7]

Subsequent to the implementation of this method, inspectors noted apparent injection sites in veal calves presented for slaughter. When analysis of the tissue from these sites was conducted and results were compared with muscle samples from other areas of these carcasses, it was found that these sites contained very high concentrations of testosterone, far in excess of reported endogenous concentrations. An additional method was therefore developed to examine tissues for the presence of hormone esters which would be indicative of external sources.[78] In the new method, which is based on the QuEChERS (Quick Easy Cheap Effective Rugged and Safe) approach,[82] samples are extracted and partitioned, filtered and then analysed using LC-MS/MS in the positive electrospray mode, using d_2-testosterone as an internal standard. Since the method targets esters of the hormones, a digestion step for bound residues is not required. The compounds included in the current method, the ions monitored, Decision Limits (CCα) and Detection Capabilities (CCβ) are given in Table 4.6. Extension of the method to additional hormone esters has been investigated. Preliminary results indicate that the extension could include a number of compounds, which have potential for non-approved use, such as boldenone undecylenate, chlortestosterone acetate and oestradiol benzoate.

Table 4.6 Compounds included in the LC-MS/MS for hormone esters and ions monitored for detection and confirmation.

Analyte	*Precursor ion (m/z)*	*Product ions (m/z)*	*CCα (μg kg⁻¹)*	*CCβ (μg kg⁻¹)*
Testosterone cypionate	413	107, 97, 109	3.1	7.8
Testosterone enanathate	401	113, 97, 109	1.7	4.2
Testosterone propionate	345	97, 109	3.4	8.5
Trenbolone acetate	313	253, 107	0.9	2.3

A recent development of a multi-class method which includes Non-Steroidal Anti-Inflammatory Drugs (NSAIDs), corticosteroids and steroidal compounds is representative of the opportunities to implement multi-class methods using the separation and detection capabilities LC/MS/MS.[83] After hydrolysis with protease, samples are initially cleaned up by liquid-liquid partitioning, followed by further clean-up using multiple SPE cartridges. The NSAIDs are retained and subsequently eluted from the second cartridge in the sequence, while the steroids and corticosteroids are eluted from the third cartridge. The two fractions are then separately analysed by LC-MS/MS. The method scope includes 10 NSAIDs, 11 corticosteroids and 8 steroids in kidney and muscle tissues, with limits of confirmation less than 1.0 μg kg^{-1} and recoveries > 50% for most compounds. Steroids detected by the method include boldenone, dianabol, nortestosterone (17α- and 17β-), testosterone (17α- and 17β-) and TREN (17α- and 17β-).

The availability of alternative methods which target some of the same compounds, plus other compounds unique to one method, enables targeted use of methods to provide the optimal application of each method in a residue control programme to a particular species or class of animal to detect those substances most likely to be used in that species/class for therapeutic or growth promotion purposes. For example, while a general method for a range of steroids may be more appropriate for feedlot beef cattle, a method that covers a broader range of substances, such as steroids, corticosteroids and NSAIDs, may be more useful in monitoring of veal. Resources required for each method can then be balanced against known or anticipated uses of substances in a particular production system.

The development of methods for hormones, particularly those that may be used in a non-approved fashion, is ongoing in many laboratories, including the regulatory laboratories in Canada. Current areas of research include the adaptation or further development of methods for urine which include both 17α-oestradiol and taleranol, access to new LC column technologies and the investigation of the use of molecularly imprinted polymers for clean up. Future adaptation of methods to other species, such as fish, may be anticipated in the future.

4.3.4.3 European Union

In the EU the development of multi-residue methods for hormones has been based mainly on the requirements set by the European Commission (EC), originally described in Council Directive 96/23/EC,[84] more recently further specified by a Community Reference Laboratory (CRL) guidance paper.[10]

A very detailed screening and confirmatory approach was described by Hewitt *et al.*,[85] who produced a semi-automated quantitative method capable of the screening and confirmation of 22 steroids in urine. Initial screening is based on LC-MS/MS. However, the LC-eluate is split and fractions are collected for further confirmatory analyses, either by further LC-MS/MS measurements (additional transitions) or GC-hrMS after derivatisation. This was one of the first methods validated based on Commission Decision 2002/657[7] and demonstrating its suitability of the $1\,\mu g\,l^{-1}$ target concentration. Kootstra *et al.*[86] also described a fully validated GC-MS method for the screening of a wide range of stilbenes and steroids, which fulfils the minimum method validation requirements of the EC. After dual SPE of C18 and Oasis™ HLB the purified extract is split in two portions, one for analysis by GC-MS as HFB-derivative, and the other as a TMS-derivative. For all analytes included in this method the values for $CC\alpha$ and $CC\beta$ are well below $1\,\mu g\,l^{-1}$. An important aspect of this procedure is the inclusion of some metabolites for strongly metabolising androgens like methyltestosterone.

Validation of analytical methods nowadays is an integral part of method development. Recent publications of new methods always contain detailed information with respect to the results of validation studies. A thorough validation protocol was used by Galarini *et al.*,[87] who described a confirmatory method for nortestosterone based on GC-MS. They especially studied the ruggedness of this method under routine conditions and concluded that the approach is fit for purpose under such conditions.

In spite of the advances that have been made using methods based on GC-MS, there has been a gradual shift from GC-MS to LC-MS/MS methods, at least for part of the compounds included in EU NRCPs. For example, Van Poucke *et al.*[88] described a method for 21 anabolic steroid residues in bovine urine based on LC-MS/MS. In this study they concluded, however, that the electrospray ionisation (ESI) mode used in LC-MS/MS was not sensitive enough for four compounds, with a hydroxyl-group on the 3 position (ethylestranediol, methandriol and methylandrostanediol) and ethynylestradiol. It was concluded that GC-MS was required for these compounds.

Based on the work previously published by Kootstra,[86] Zoontjes *et al.* modified this procedure for the matrix meat and the use of LC-MS/MS.[89] This method comprised the screening of both androgens and oestrogens at a concentration of $0.6\,\mu g\,kg^{-1}$. This method has been implemented in several EU laboratories and has demonstrated robustness during transfer to other laboratories.

An extensive review of recently published multi-residue chromatographic methods for the determination of steroid hormones in edible tissues was published in 2008 by Noppe *et al.*[90] This paper gives an overview of sample extraction and purification techniques as they have been employed over recent years, describing the use of supercritical fluid extraction (SFE), solid phase microextraction (SPME), molecularly imprinted polymers (MIPs) and size exclusion chromatography (SEC/GPC). Gradually, these techniques have replaced procedures based on solvent extraction and liquid-liquid partitioning. Nevertheless, modern sorbents available for SPE-based methods still make this the most frequently used approach. The authors conclude that chromatographic separation methods (GC or LC) coupled to sensitive and specific detection systems such as MS dominate the determination of steroid hormones in edible tissues. In addition, for the future, advanced techniques like ToF-MS and cyclotron resonance (ICR) and Orbitrap MS might become very useful techniques in identifying previously unknown compounds in biological samples.

In 2008 a detailed study was also made by van Rossum *et al.*[36] in which a procedure as described above for the stilbenes was validated, specifically for the power to confirm the identity of a wider range of hormonal compounds at low concentrations using GC-MS/MS. Table 4.7 lists the compounds and the ions used for screening and confirmation in this study. Since the objective of this method is to analyse samples for a broad range of compounds, SPE clean-up is used instead of LC-fractionation. In this particular case the SPE was performed using a Varian SPE C18 column. The SPE column was washed with, respectively, water and methanol/water (40/60 v/v) and elution of analytes was achieved with methanol/water (80/20 v/v). After further clean-up using a liquid-liquid extraction with pentane the dried extract was reconstituted and derivatised with MSTFA^{++}. GC-MS/MS was carried out on 30 m VF-17MS (Varian) i.d. 0.25 mm, 0.25 μm film thickness column with temperature ramp between 110 °C and 340 °C.

This multi-residue method for muscle was validated according to EU Commission Decision 2002/657/EC[7] and for most compounds confirmation at 1.0 μg kg^{-1} was possible. In a previous study, GC-lrMS (GC-MSD unit mass resolution) was evaluated for a limited number of compounds, but in samples obtained from different species.[68] The targeted clean up used in the GC-lrMS method was replaced in the new GC-MS/MS method with a clean-up procedure that was both generic and quicker to perform. This in turn permitted a much larger number of analytes to be included, over 20 compared to 4 in the original procedure. It must be stated, though, that the percentage of samples in which the identity was fully confirmed, based on the four IP criteria, was similar in the two studies, approximately 50% at 0.5 μg kg^{-1} (Table 4.8), which is one half the currently recommended concentration for the majority of hormones in muscle (Table 4.2[10]). These results show that the generic clean-up approach, combined with more advanced (GC-MS/MS) detection, gives similar results when compared to extensive and compound specific clean-up procedures used with GC-lrMS. However, the strength of the generic GC-MS/MS approach is the extended range of compounds that can be included during routine analyses, which is often of more importance than improved limits of detection limits/confirmatory power when used as a screening technique.

Table 4.7 Parameters GC-MS/MS (the collision energy is given in brackets). Grey fields represent internal standards.

Analyte	*Retention time (minutes)*	*MRM1 (m/z)(screening)*	*MRM2 (m/z) (confirmation)*
37-chloromadinone	11.89	580 > 231 (–5V)	580 > 490 (–10V)
Benzoestrol	9.82	207 > 179 (–10V)	207 > 191 (–10V)
Chloromadinone	11.89	578 > 231 (–15V)	Not present
Chlorotestosterone	11.43	466 > 335 (–15V)	466 > 431 (–10V)
Chlorotestosterone-acetate	12.33	436 > 401 (–15V)	436 > 230 (–20V)
Chlorotestoster-oneacetate-d3	12.32	439 > 404 (–5V)	
Chlortestosterone-d3	11.43	469 > 338 (–5V)	
cis-diethylstilboestrol	8.64	412 > 217 (–20V)	412 > 383 (–15V)
cis-diethylstilboestrol-d6	8.62	418 > 220 (–25V)	
Dienoestrol	9.21	410 > 381 (–5V)	410 > 395 (–5V)
Dienoestrol-d2	9.21	412 > 397 (–20V)	
Ethynylestradiol	11.10	425 > 231 (–14V)	425 > 205 (–14V)
Ethynylestradiol-d4	11.09	429 > 233 (–15V)	
Hexoestrol	8.99	207 > 179 (–10V)	207 > 191 (–10V)
Hexoestrol-d4	8.98	209 > 180 (–5V)	
Medroxyprogsterone	11.32	560 > 328 (–15V)	560 > 315 (–15V)
Medroxyprogsterone-d3	11.31	563 > 331 (–25V)	
Megestrol	11.29	453 > 273 (–17V)	Not present
Megestrol-d3	11.29	456 > 276 (–15V)	
Melengestrol-d3	11.33	573 > 483 (–10V)	
Melengestrol	11.34	570 > 480 (–10V)	570 > 465 (–15V)
Methylboldenone	10.70	444 > 206 (–15V)	444 > 339 (–10V)
Methyltestosterone	10.70	446 > 301 (–30V)	446 > 356 (–5V)
Methyltestosterone-d3	10.69	449 > 301 (–30V)	
Norclostebol	11.39	452 > 417 (–5V)	452 > 321 (–5V)
Norclostebol-acetate	12.28	422 > 216 (–10V)	422 > 387 (–5V)
Norethandrolone	11.11	446 > 356 (–10V)	446 > 287 (–20V)
Normethandrolone	10.62	432 > 287 (–25V)	432 > 342 (–15V)
Progesterone	11.33	458 > 443 (–5V)	458 > 157 (–20V)
Progesterone-d5	11.31	463 > 448 (–15V)	
β-boldenone	10.35	430 > 206 (–18V)	430 > 325 (–12V)
β-boldenone-d3	10.34	433 > 206 (–15V)	
β-oestradiol	10.60	416 > 285 (–16V)	416 > 326 (–18V)
β-oestradiol-d3	10.59	419 > 285 (–28V)	
β-nortestosterone	10.26	418 > 313 (–12V)	418 > 328 (–10V)
β-nortestosterone-d3	10.25	421 > 316 (–20V)	
β-testosterone	10.35	432 > 209 (–10V)	432 > 327 (–5V)
β-testosterone-d2	10.35	434 > 211 (–11V)	
Trans-diethylstilboestrol	9.06	412 > 217 (–20V)	412 > 383 (–15V)
Trans-diethyl-stilboestrol-d6	9.04	418 > 220 (–25V)	

The instrumental techniques discussed so far have employed either GC or LC chromatography coupled to mass spectrometry, although it is apparent from recent publications that LC-MS/MS using triple Quadrupole (QqQ) type systems is fast becoming the technique of choice for the detection of hormones.

Table 4.8 Overview of confirmation analysis for meat samples spiked with different concentrations of compounds (μg kg^{-1}). Grey shaded fields are confirmed according to EU Commission Decision 2002/657/EC.

	Ion ratio limits from standards (m/z No.1)/(m/z No.2)	*Measured ion ratio for spiked samples of meat (μg kg^{-1})*				
Analyte	*Range (minimum–maximum)*	*0.25*	*0.5*	*0.75*	*1.0*	*2.5*
Benzestrol	20.1–33.4	68.3	82.0	59.9	46.4	31.0
Chloromadinone	24.6–41.1	11.3	92.2	34.2	39.1	81.1
Chlorotestosterone	31.0–51.7	40.3	39.9	43.3	43.4	38.2
Chlorotestosterone-acetate	12.2–22.7	16.7	43.0	43.8	13.9	20.2
Dienoestrol	23.0–38.4	31.6	26.2	24.7	30.6	28.4
Diethylstilboestrol	64.5–96.8	72.7	70.5	54.0	65.9	91.8
Ethynylestradiol	33.5–55.9	45.1	34.1	33.7	35.0	45.3
Hexoestrol	20.7–34.4	30.6	28.4	27.9	31.8	26.3
Medroxyprogesterone	22.3–37.2	84.5	4.0	9.9	28.7	20.8
Methylboldenone	26.1–43.4	77.8	22.8	16.6	21.5	31.1
Methyltestosterone	46.5–69.8	64.0	114	111	78.1	83.0
Norclostebol	77.2–116	38.6	78.2	68.5	250	91.5
Norclostebol-acetate	75.2–113	51.5	114	84.8	102	101
Norethandrolone	35.0–58.3	23.9	45.2	47.4	33.2	58.4
Normethandrolone	24.5–40.8	36.2	13.3	27.6	27.7	33.6
Progesterone	40.3–60.4	105.4	153.6	141.16	130.3	130.6
β-boldenone	23.8–39.7	48.8	33.5	32.8	19.5	31.0
β-oestradiol	16.6–27.7	38.1	34.5	13.8	15.7	20.1
β-nortestosterone	75.4–113	128	136	176	115	112
β-testosterone	6.1–18.3	4.6	18.9	12.6	13.3	9.1
Percentage of analytes with confirmed ion ratios		50	40	50	60	85

There are several reasons for using LC- and not GC-MS. One of the most important is that many hormones are relatively polar compounds which need derivatisation prior to GC-analyses. Moreover, one of the problems residue chemists have to face is metabolism. After administration these compounds undergo extensive phase I and II metabolism, in which they are converted to more polar compounds and are excreted in urine. The main phase I metabolic pathways are oxidation, hydrolysis and reduction, which often bring a more polar group to the steroid structure, offering a site for the conjugation in phase II metabolic reactions. The most common phase II conjugation reaction in animals is glucuronidation in which the steroid is coupled to glucuronic acid. Development of alternative direct analysis methods for steroid-conjugates is thus of great importance. Liquid chromatography-mass spectrometry using electrospray ionisation (LC-ESI-MS) is a suitable approach for such analyses. Study of the mass spectrometric behaviour of anabolic steroid-glucuronides and -sulfates is essential in the development of direct methods of analysis by LC-MS. To this time a limited number of MS studies have been published with most work presented in literature focused on the analysis of (academic) standards. Wubs *et al.*[91] studied the anabolic steroid-conjugates presented in Table 4.9. The selection of the compounds represents structurally interesting steroids having slight differences in the substitution at carbons 3, 5, 10 and 17.

In this method conjugated steroids were isolated from urine using an OASIS™ HLB SPE cartridge. The LC was carried out *via* a Waters Chromatography Acquity UPLC separation module using a BEH C18 1.7 µm (100 × 2.1 mm ID) column with MS analysis in negative ESI mode. Table 4.9 also lists the measured MRM transitions. These experiments determined that the fragmentation of conjugated steroids obtained in tandem mass spectrometry (MS/MS) electrospray (ESI) positive ion mode was more specific than in ESI negative mode. Since, however, the analyte sensitivity was found to be much higher in negative ion mode, this approach was used to study both the glucuronide- and sulfate-steroids using one LC-MS/MS method. Figure 4.1 shows a chromatogram from a sample containing a mixture of glucuronide- and sulfate-conjugates spiked to bovine urine. Each trace represents the measured transition for the given compound. In all cases, the origin of the product measured is from the glucuronide moiety or sulfate moiety. The transitions with the highest relative abundance were chosen. In Table 4.10 an overview is given of the validation results from this method, in terms of CCα and CCβ.

Due to high background concentration of some of these compounds in the urine test samples it was not possible to determine CCα and CCβ in the traditional way. To overcome this, additional water samples were spiked and processed. This approach gives a more realistic estimate of the CCα and CCβ, which, as can be expected, results in lower values for all compounds. For all compounds the CCα is equal to or lower than 2.2 $\mu g\, l^{-1}$. As a rule the values for urine are much higher due to the high background concentrations present in the materials analysed. For example the high background for boldenone-glucuronide is caused by the presence of 17α-boldenone in the samples analysed. It is well established that this compound can be present in samples of bovine urine

Table 4.9 Steroid glucuronides and sulfates and their corresponding MRMs.

Name	*Abbreviation*	*Formula*	*MRM*	*Supplier*
Androsterone-d5-glucuronide	And-d5-glu	$C_{25}H_{33}D_5O_8$	470.3 > 85.2	DSHS (Germany)
Androsterone-glucuronide	And-glu	$C_{25}H_{38}O_8$	465.2 > 85.2	Ikapharm
Boldenone-d3-sulfate (Na-salt)	Bold-d3-su	$C_{19}H_{22}D_3O_5S$	368.2 > 353.2	NARL (Australia)
Boldenone-glucuronide	Bold-glu	$C_{25}H_{34}O_8$	461.2 > 113.2	Rikilt (Netherlands)
Boldenone-sulfate (Na-salt)	Bold-su	$C_{19}H_{26}O_5S$	365.2 > 350.2	NARL (Australia)
DHEA-glucuronide	DHEA-glu	$C_{25}H_{36}O_8$	463.2 > 113.2	Sigma
DHEA-sulfate (Na-salt)	DHEA-su	$C_{19}H_{28}O_5S$	367.2 > 97.2	Schering
Oestradiol-glucuronide (Na-salt)	E2-glu	$C_{24}H_{32}O_8$	447.2 > 85.2	Sigma
Oestradiol-sulfate	E2-su	$C_{18}H_{24}O_5S$	351.2 > 271.2	Sigma
Oestrone-d4-sulfate (Na-salt)	E1-d4-su	$C_{18}D_4H_{18}O_5S$	353.2 > 273.2	Sigma
Oestrone-glucuronide (Na-salt)	E1-glu	$C_{24}H_{30}O_8$	445.2 > 113.2	Sigma
Oestrone-sulfate (K-salt)	E1-su	$C_{18}H_{21}O_5S$	349.2 > 269.2	Sigma
Pregnenolone-sulfate (Na-salt)	Preg-su	$C_{21}H_{31}O_5S$	395.2 > 97.2	Sigma
Progesterone-glucuronide	Prog-glu	$C_{27}H_{38}O_9$	505.2 > 113.2	Sigma
Testosterone-glucuronide	T-glu	$C_{25}H_{36}O_8$	463.2 > 113.2	Sigma

obtained from untreated animals.[1] The current LC-system used in this method is, however, not suitable for separating the different conjugated epimers. Future applications for the analyses of conjugated steroids are mainly expected in the area of discriminating the natural presence and presence due to abuse of natural hormones. The analysis for 17β-boldenone glucuronide is one example of the usefulness of this approach. However, method sensitivity and robustness still require further improvement.

4.3.4.4 Australasia

In Australia the National Residue Testing Programmes for hormones are divided into two programmes[92] operated by contract laboratories. The first programme includes stilbenes, resorcylic acid lactones (RALs) and TREN. The second programme covers androgenic hormones. Most methods are based on published methods with the laboratories making changes to suit their circumstances.

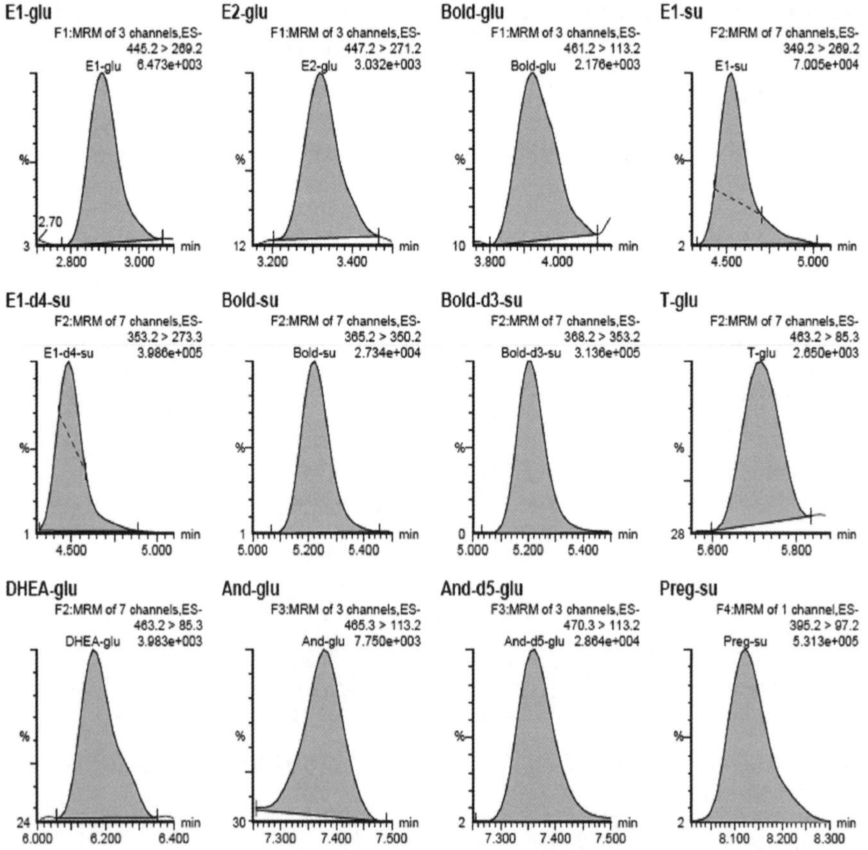

Figure 4.1 Reversed-phase microbore LC-ESI MSMS profiles of a sample of urine spiked ($5\,ng\,ml^{-1}$) with a mixture of steroid-glucuronide- and sulfate-conjugates (smoothed mean 3).

Table 4.10 Validation results for the direct LC-MS analysis of steroid glucuronides and sulfates, CCα and CCβ

	CCα (μg l⁻¹)		*CCβ (μg l⁻¹)*	
Analyte	*Spiked water*	*Urine*	*Spiked water*	*Urine*
E1-glu	1.2	2.4	1.7	3.4
E2-glu	2.2	5.9	3.1	8.5
Bold-glu	1.5	6.2	2.1	8.8
E1-su	0.7	2.2	1.0	3.1
T-glu	0.6	1.7	0.9	2.5
DHEA-glu	0.8	40.6	1.2	57.7
Bold-su	0.2	0.4	0.3	0.6
Prog-glu	0.8	9.9	1.2	14.2
And-glu	0.8	6.5	1.2	9.3
Preg-su	1.4	15.3	1.9	21.8

GC-MS analysis is used for stilbenes and RALs following hydrolysis with β-glucuronidase, solvent extraction (hexane, dichloromethane/acetonitrile), clean-up using ion exchange chromatography and trimethylsilyl derivatisation. TREN is analysed using a separate UPLC-MS/MS method following hydrolysis with β-glucuronidase, extraction with TBME and a SPE clean-up. Additional confirmation is by HRGC/HRMS. Quantification of analytes employs deuterated analogues of each target analyte except for zearalenone and dienoestrol. Zearalenone is quantified based on deuterated zeranol and dienoestrol based on deuterated DES. The quoted Limits of Determination (LODs) and Limits of Reporting (LORs) for dienoestrol, DES and HEX are 0.1 $\mu g\, kg^{-1}$ and 0.2 $\mu g\, kg^{-1}$, respectively. The LOD and LOR for α-zearalanol ZER, β-zearalanol (taleranol), α-zearalenol, β-zearalenol, zearalenone, zearalanone and TREN are 1 $\mu g\, kg^{-1}$ and 2 $\mu g\, kg^{-1}$. Confirmation of identity meets EU identification criteria as outlined in EU Commission Decision 2002/657/EC.[7] These methods cover liver, urine and faeces.

For androgenic hormones, GC-MS is used for methandriol (methylandrostenediol) analysis. This procedure includes β-glucuronidase hydrolysis of samples and C18 SPE clean-up followed by acid hydrolysis, solvent extraction and derivatisation with t-butyldimethylsilyl. The other steroids are cleaned up after acid hydrolysis using C18 SPE rather than solvent extraction and analysed using LC-ESI-MS/MS. Quantification of nortestosterone, boldenone and stanozolol is based on deuterated analogues of nortestosterone, boldenone and stanozolol respectively. Methandriol and 16-OH stanozolol are quantified based on external standards. The analytes covered and LODs/LORs are: 16-OH stanozolol, stanozolol, 17α-nortestosterone, 17β-nortestosterone, 17α-boldenone and 17β-boldenone with a LOD and LOR for all set at 1 $\mu g\, l^{-1}$. The LOD for methandriol is 1 $\mu g\, l^{-1}$ and the LOR is 5 $\mu g\, l^{-1}$. Confirmation of identity meets EU identification criteria as outlined in EU Commission Decision 2002/657/EC[7] and urine is the preferred test matrix.

4.3.5 Use of LC-ToF-MS as a Multi-residue Screening Method

4.3.5.1 Current Situation

The methods described above all have one principle thing in common: they were all developed for the analyses of a limited number of compounds. In spite of the fact that most methods are multi-residue methods, there always is a limitation in the number of analytes. Primary causes for these limitations are poor full-scan sensitivity and specificity of the LC-MS/(MS) equipment. An attractive alternative for LC-MS/MS is the use of full mass scan MS techniques, for example Time-of-Flight (ToF). The medium to high resolution of 10,000 FWHM of the ToF provides a significant selectivity and therefore sensitivity gain compared to unit-resolution scanning MS instrumentation. A significant advantage of ToF-MS is that no *a priori* hypothesis about the presence of certain drugs is required; that is, no analyte-specific transitions have to be defined before injecting the sample. The high-resolution, full scan data permit the testing of any *a posteriori* hypotheses by extracting any desired exact mass chromatogram. Moreover, the accurate mass capability of LC-ToF-MS allows the reconstruction of highly selective accurate mass chromatograms of target residues in complex matrices, for example the simultaneous determination of different groups of antibiotic compounds in milk.[93]

LC-ToF-MS is sensitive, as well as yielding specific results (exact mass) in full scan measurements. These mass spectrometers can be used to detect multiple compounds in one extract. However, other factors should also be considered:

- Sample clean-up should not be too specific. The matrix has to be stripped of interfering compounds.
- LC separation should be capable of performing excellent chromatography for a large number of compounds.
- Using an ESI/APCI combination source makes it possible to ionise most of the target compounds.
- Identification software and library search algorithms must be able to determine the identity of the compound.

The development of such a method has been described by Zoontjes *et al.*[94] A list of compounds was selected covering a broad range in the field of residue analysis and a range of different matrices was selected. In Figure 4.2 a schematic overview is given of the matrix/analyte combinations tested in the developed method.

In the case of muscle samples the primary extraction was performed using ultrasonic disruption, TBME extraction and defatting after dissolving the dried

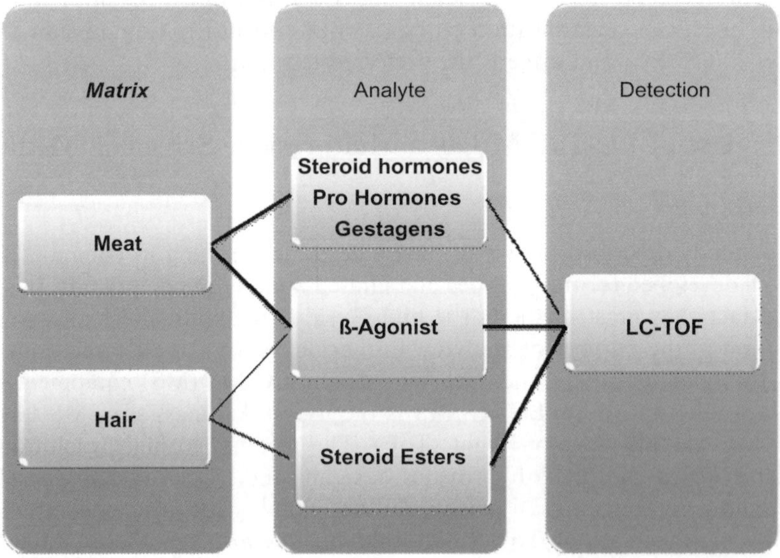

Figure 4.2 Overview of matrix–compound combinations tested.

primary extract in methanol/water. The defatted extract was further purified by C18 SPE prior to determination by LC-ToF-MS (Agilent 6210 LC). Analyte separation was *via* a Zorbax Rapid Resolution C18 (2.1 × 50 mm, df=1.8 µm) column. For detecting steroid esters in hair, samples are washed, cut and homogenised before analysis. Methanol is used as the extraction solvent, after addition of internal standard (testosterone-phenyl-propionate). Sample clean-up is *via* SPE (Baker Bond Octadecyl).

Currently MS analysis by LC-ToF-MS is not described in EU Commission Decision 2002/657/EC.[7] Consequently, no criteria are set for confirmation using this type of instrument. There are, however, validation criteria concerning the qualitative aspect of the method, which can be used. Qualitative methods can be validated by analysing 20 samples spiked with the compounds of interest and 20 blank samples. From the spiked samples at least 95% should be non-compliant and for the blank samples no false non-compliant results should be obtained. For compounds included in this study, the mono-isotopic mass for each compound was extracted plus/minus 10 ppm. A compound was stated identified when the S/N of the signal at the expected retention time exceeded three.

Table 4.11 provides an overview of the validation results, with the theoretical detection limits given in the third column. In the case of 100 percent recovery and no interference of the signal, by for example ion suppression, the compounds should be detectable at these concentrations. The clean-up methods described were originally developed and validated for other LC-MS/MS methods and the recovery was estimated from these methods. After analysis, the samples were processed using automated data processing software. In the last column the percentage of non-compliant samples at the spiking concentration is given.

These results show that LC-ToF-MS is potentially a significant improvement over more traditional approaches. The main reason for this is that the high scanning speed allows the very sensitive detection of a large number of analytes over a short period of time. Nevertheless, there is still room for improvement. From the last column it can be seen that the analyte identity was not confirmed in all cases.

4.3.5.2 Future Potential of the ToF-MS Technique

Since the publication of the method for DES in 1986 there have been various and significant improvements made in the tools for sample clean-up, for chromatography and for mass spectrometry that are available to residue chemists. These tools have paved the way for continued work on generic (non-targeted) techniques. Elsewhere in this book, work on biosensor receptors and whole cell lines have been discussed. These methods also have the objective to identify any compound, which fulfils certain criteria, for example hormonal activity comparable to testosterone (androgenic activity) or oestradiol (oestrogenic activity). In part, these procedures are already successful, but issues

Table 4.11 Overview of the validation results.

	Mono-isotope mass	*Theoretical detection limit ($\mu g\,kg^{-1}$)*	*Spike concentration ($\mu g\,kg^{-1}$)*	*Non-compliant (%)*
MEAT STEROIDS				
Clostebol	323.1772	>20	4	0
Methylboldenone	301.2162	0.8	2	50
Mibolerone	303.2319	0.8	4	10
Progesterone	315.2319	0.8	4	75
β-nortestosterone	275.2006	0.8	2	95
β-testosterone	289.2163	0.8	2	85
β-trenbolone	271.1693	0.8	1	95
Stanozolol	329.2588	0.8	1	20
MEAT PRO-HORMONES				
1,(5α)-Androstene-17β-ol-3-one	289.2163	0.8	2	95
1,4-androstedienedione	285.1850	1.25	2	50
4-androsten-17α-ol-3-one	289.2163	0.8	2	50
5-androsten-17β-ol-3-one	289.2163	1.25	2	50
5-androsten-3,17-dione	287.2006	1.25	2	35
MEAT GESTAGENS				
Chlormadinone acetate	405.1828	1.7	2	0
Delmadione acetate	403.1671	1.7	2	100
Fluoxymesterone	337.2174	1.7	2	70

Flurogestone acetate	407.2229	1.7	2	70
Megestrol acetate	385.2374	1.7	2	100
Melengestrol acetate	397.2374	1.7	2	100
HAIR STEROID ESTERS				
Dehydrotestosterone-un-decylenate	453.3283	30	160	100
Oestradiol-17-propionate	329.2112	40	160	100
Oestradiol-acetate-17β	315.1955	10	160	100
Oestradiol-valerate	357.2425	30	160	100
Norclostebol acetate	351.1803	40	160	50
Nortestosterone-acetate	317.2112	10	160	100
Nortestosterone-laurate	457.3677	160	160	50
Testosterone-cypionate	413.3051	10	160	100
Testosterone-acetate	331.2268	5	160	75
Testosterone-decanoate	443.3520	10	160	0
Testosterone-enanthate	401.3051	10	160	100
Testosterone-phenyl-propionate	421.2738	10	160	100
Testosterone-propionate	345.2425	5	160	100

remain. For example the detection of prohormones, not yet possessing biological activity, or inactivated metabolites, is a complicated topic. Further, these bioassays have one drawback; they do not identify the compound responsible for the biochemical response. Therefore, there also is a need for methods that can identify such compounds. The use of LC-ToF-MS is a possible approach, but very specific conditions are necessary. One of those is the use of more intelligent data evaluation procedures. The systematic evaluation of specific sets of ToF-MS data in combination with statistical evaluation potentially makes it possible to discriminate between samples obtained from animals treated with *e.g.* a hormone and untreated (control) animals. Figure 4.3 presents an example of such a data evaluation approach.

In this study, 20 meat samples fortified with 2 $\mu g\, kg^{-1}$ of a steroid (MGA) and 20 blank meat samples were analysed using a generic extraction procedure followed by LC-ToF MS. The full scan data (50–1000 amu) of each sample were stripped from (chemical) noise (see Figure 4.3) and all detected peaks were identified by using the Agilent Mass Hunter Profiling software. The combined information for all samples was systematically evaluated and the differences – if tested statistical significant – were plotted using the Agilent Mass Profiler software. Figure 4.4 presents the abundances (log2) of the peaks detected in the blank samples *versus* the abundances (log2) of the peaks detected in the spiked samples. When a peak is beyond the four-fold margin (demonstrated by the lines in the data plot) this peak is marked as nearly unique for that group. In this example one signal is more abundant in the control group whereas six signals are identified as more significantly present within the group of spiked samples. The strength of (Q)-ToF-MS is that the accurate mass and/or the spectra of the deviating signals can be reproduced and, through suitable library searching, identified. In this case, one of the compounds, represented as a circled spot in Figure 4.4, was identified as MGA. Much work however remains to be done; in the areas of developing both generic clean-up procedures, LC-ToF-MS parameters, and data evaluation software. Moreover, the accurate definition of the reference population of signals will be very critical and very different for each species/matrix. Further development is thus required to make this approach generally applicable, but potentially it could revolutionise the control for banned substances, not only in veterinary practices, but also in related fields like sports-doping analyses.

4.4 Zeranol – a Special Case Regarding Mycotoxins

As discussed in the introduction to this chapter there are often a number of external factors to be considered during final data analysis. One of the most striking examples of this is zeranol (a semi-synthetic oestrogenic growth promoter) that was banned in the EU in 1988 but continues to be used in other parts of the world *e.g.* North America. Whilst it was widely known that zeranol residues could appear in urine and animal tissues after the administration of zeranol, a number of researchers in Europe and New Zealand were also

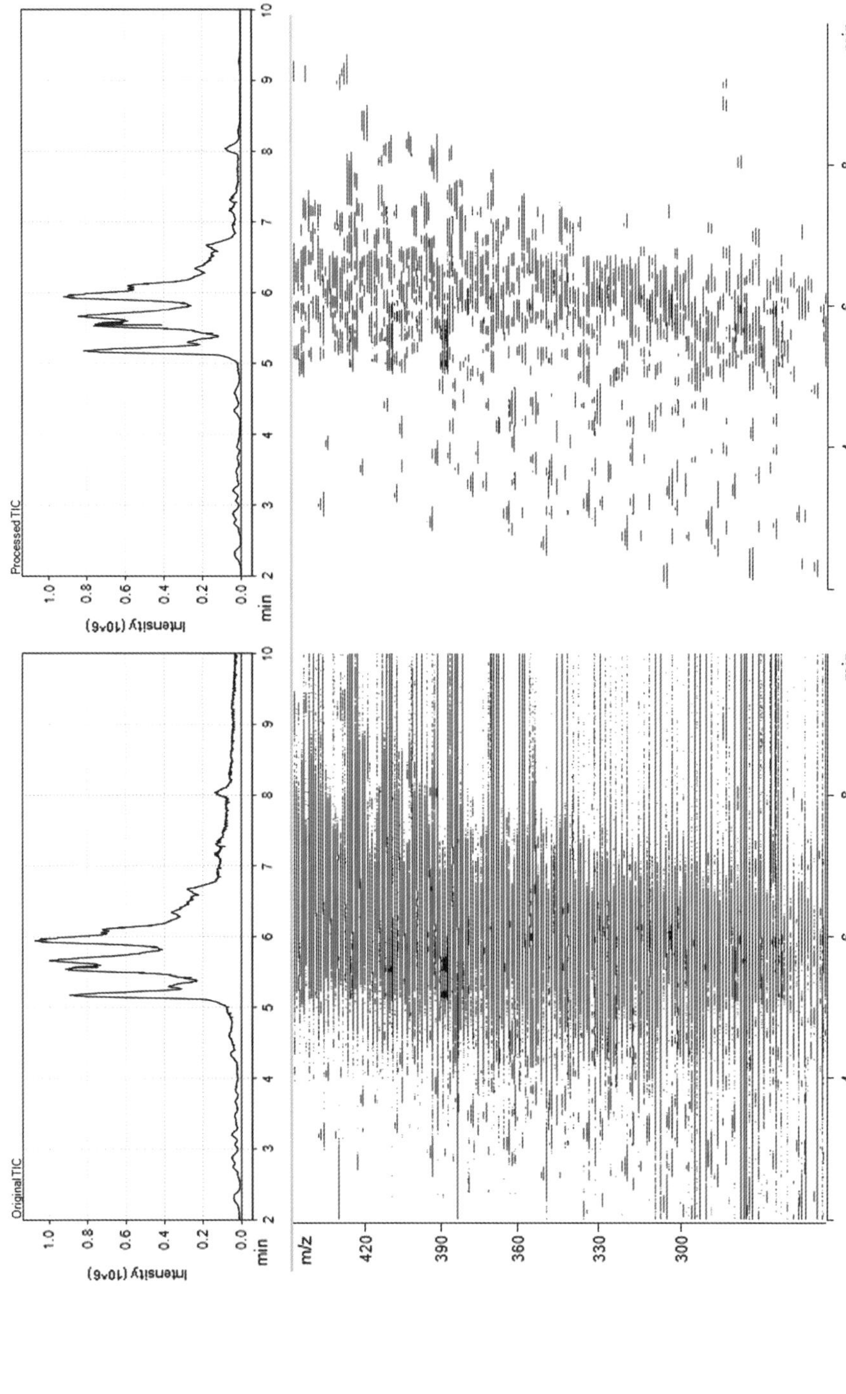

Figure 4.3 Contour plots showing the raw data from a meat extract (left) and the stripped data (right).

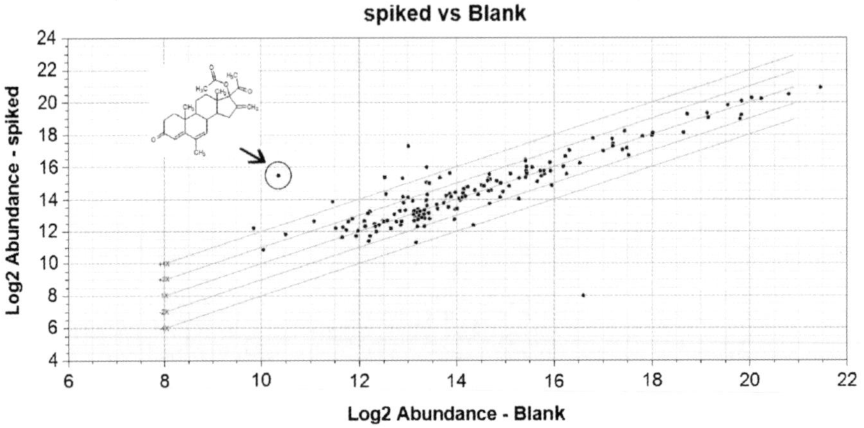

Figure 4.4 Melengestrol acetate is uniquely identified in the group of spiked samples.

concerned about its possible presence in test samples due to *in vivo* conversion of *Fusarium* spp. toxins present in animal feed. In 1998 Kennedy *et al.*[95] demonstrated that zeranol could be formed *in vivo* and detected in bovine bile following the oral administration of α-zearalenol and zearalenone, but not following administration of β-zearalenol. This study also showed that, whilst *Fusarium* spp. toxins were present in 32% (n=422) of all bovine bile tested, zeranol was only confirmed in 6.6% of cases (n=28). Mean α- and β-zearalenol concentrations in the bile of zeranol positive animals were typically much higher (*c.* 10 times) than those in the negative control samples. It was postulated that, since the α-zearalenol concentration was always at least five times higher than the zeranol concentration, this could be used as a means to discriminate between zeranol abuse (in the EU) and natural contamination. To assist with this work a rapid screening method based on time resolved – fluoroimmunoassay (TR-FIA) – was also developed for zeranol and α-zearalenol[96] and subject to an inter-laboratory comparison with four commercially available ELISAs for zeranol.[97] The claimed advantage of the TR-FIA over the four ELISAs was that fewer "false positive" results were obtained. This conclusion was based upon the fact the TR-FIA was highly specific for zeranol and was free of interferences (no false positive results) from naturally occurring *Fusarium* spp. toxins. In contrast, when zeranol-free incurred samples containing *Fusarium* spp. toxins were analysed by the commercial ELISAs, three out of the four kits produced a significant number of false-positive results.

The validated TR-FIA method was subsequently employed by Launay *et al.*[98] in a survey of over 8000 urine samples collected from four EU control laboratories. Of the samples tested *c.* 94% tested negative for zeranol and all screening positive samples were then re-analysed by a confirmatory method based on either GC-MS or LC-MS/MS. A linear regression comparison of the screening *versus* confirmatory data (using a 99% confidence interval) revealed that 170 out of the 174 suspect samples belonged to a normal population

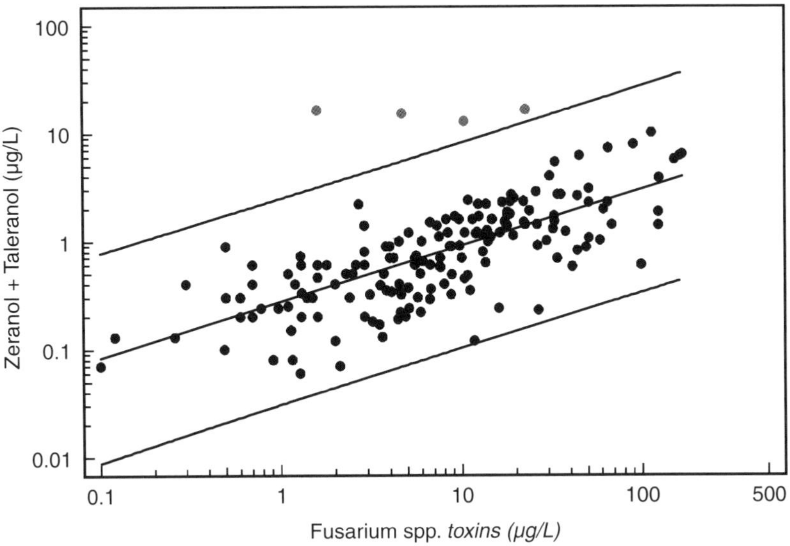

Figure 4.5 Samples (n=174) belonging to the suspect sample population which contained zeranol and taleranol in addition to *Fusarium* spp. toxins. The samples (n=4) that fell outside the 99% confidence interval are outliers and may be indicative of zeranol abuse. Reproduced from reference 97.

whereby the amount of zeranol (and taleranol) could be shown to have a direct relationship with the total amount of *Fusarium* spp. toxins in the sample (Figure 4.5). The remaining four samples were attributed to possible zeranol abuse rather than to natural contamination *i.e.* truly "positive" samples.

4.5 Performance of Current Methods in Proficiency Tests

Proficiency testing has been an ongoing part of the residue control programmes in North America since the early beginnings of the residue control programmes for veterinary drug residues in the United States of America and Canada, with a regular exchange of materials for sulfonamide drug residues dating to the early 1980s. While the original focus for USDA/FSIS laboratories and their Canadian counterparts was on sulfonamides and chlorinated pesticides, exchanges have included a variety of other drugs, including antimicrobials and hormones. In 2001, the Canadian Food Inspection Agency established a separate proficiency testing unit at the Centre for Veterinary Drug Residues, located at the CFIA Saskatoon Laboratory. This PT Unit provides 3–4 rounds of PT samples each year to a group of primarily Canadian residue control laboratories, including federal, provincial and private laboratories engaged in the analysis of samples for federal and provincial inspection programmes. The PT samples are used to support the activities of the Standards Council of

Canada in accrediting these laboratories under ISO-17025 for specified tests, including a range of hormones, and to provide ongoing evidence of the capabilities of the laboratories involved.

Overall, performance of laboratories in the PT rounds has been satisfactory, as evidenced (in part) by the scope of accreditation listed for each participating laboratory, which may be accessed on the website of the Standards Council of Canada. The focus of the PT rounds is on matrices which are targeted in the National Chemical Residue Monitoring Programme in Canada, usually either muscle, fat, liver or kidney. To be meaningful, it is considered that the PT materials should meet certain criteria, including being representative of typical matrices which are routinely analysed by the participants, use of analyte/matrix combinations of established stability and inclusion of concentrations which reflect the established performance range of the test method. When only a small number of laboratories (3–5) participate in a proficiency round, professional judgement becomes an important factor in the interpretation of results. Reliance on a formula-based statistical approach in such instances can lead to questionable conclusions.

Incurred TREN materials produced by CFIA have also been provided to FAPAS® for use in one of their proficiency rounds. A PT advisory committee which includes representatives of CFIA, other laboratories involved in the PT rounds and independent experts has recently been formed to provide scientific guidance. Participants in the PT programme receive confidential reports on each round, plus an annual report on overall performance.

Within Europe PT-programmes for hormones are primarily organised within the CRL/NRL (National Reference Laboratory) network of laboratories. Approximately twice a year a PT is organised by the CRL. The objectives are two-fold. Regular PT is primarily organised for laboratories to demonstrate their skills with respect to residue analyses of analyte/matrix combinations considered important. NRLs also need to demonstrate good performance in order to maintain their ISO-17025 accreditation. In the latest study on 17β-oestradiol, sets of three samples of lyophilised bovine urine and plasma were distributed among 15 participating laboratories. The majority of participants reported good results for both matrices, and at a concentration of approximately 0.1 $\mu g\,l^{-1}$ in plasma. The most recent study finalised was on medroxyprogesterone acetate in porcine kidney fat. In this study laboratories were asked to quantify the mass concentration and to confirm the identity of the analyte. Target concentrations were 0, 1 and 4 $\mu g\,kg^{-1}$. The performance of the laboratories was scored on the basis of the agreement between its result and the overall average value and the ability to confirm the identity. Of the 28 laboratories 50% scored the maximum number of 9 points to be obtained, and over 90% had an acceptable score. This type of PT has proven to be highly useful for laboratories to demonstrate or to improve their performance.

Within this EU CRL/NRL laboratory network there is, however, a continued interest in further extending the methods to other compounds, *e.g.* marker metabolites and other matrices. Therefore, approximately one out of three PTs are organised as research studies. Again, all NRLs are invited to participate, but the type of analysis does not necessarily belong to the regular activities of the surveillance laboratories. Recent examples include research-PT for urine analyses after a treatment of bovine animals with pro-hormones, hair analyses, and the testing of urine for metabolites of methyl testosterone. Based on these results, laboratories have the option to include this new type of testing in their national testing programme.

4.6 Conclusions

Confirmatory analyses using instrumental techniques have made enormous progress over the past two to three decades. This progress is not so much in the concentration of hormones that can be detected and confirmed, but more in (i) the robustness of the techniques, (ii) the scope of hormones that can be detected and (iii) flexibility in matrices that can be analysed. The improved effectiveness of residue control programmes is a perhaps a better judge of progress than the Limit of Detection that can be obtained.

On the other hand, most methods currently used focus on the detection, quantification and confirmation of the identity of a limited number of compounds or classes of compounds. Especially in the area of banned substances this is a major drawback. Those who seek use of these compounds for their own profit will try to use new compounds not included in residue control programmes. Until now this means that residue chemists and legislators are always one step (at least) behind. Tools are, however, now being developed to solve this problem by developing generic (non-targeted) methods of analyses, capable of detecting deviations in patterns derived from the chemical composition of a sample. These developments will take place on the basis of biochemical techniques as well as instrumental techniques. The next five years will show whether these two combined approaches will achieve their ultimate goal: the development of generic control programmes suitable for the large-scale screening of samples using a biochemical method, combined with generic confirmatory techniques for identification of the compound(s) responsible for the response in screening.

References

1. H. F. De Brabander, S. Poelmans, R. Schilt, R. W. Stephany, B. Le Bizec, R. Draisci, S. S. Sterk, L. A. van Ginkel, D. Courtheyn, N. van Hoof, A. Macri and K. De Wasch, *Food Addit. Contam.*, 2004, **21**, 515.
2. B. Le Bizec, F. Courant, I. Gaudin, E. Bichon, B. Destrez, R. Schilt, R. Draisci, F. Monteau and F. André, *Steroids*, 2006, **71**, 1078.
3. M. H. Blokland, H. J. van Rossum, S. S. Sterk, L. A. van Ginkel and R. W. Stephany, *Anal. Chim. Acta*, 2007, **586**, 147.
4. H. H. Meyer, D. Falckenberg, T. Janowski, M. Rapp, E. F. Rösel, L. van Look and H. Karg, *Acta Endocrinol (Copenh).*, 1992, **126**, 369.

5. J. D. McEvoy, W. J. McCaughey, W. J. Cooper, D. G. Kennedy and B. M. Carten, *Vet. Quart.*, 1999, **21**, 8.
6. D. G. Kennedy, H. D. Short, S. R. Vrooks, P. B.Young, H. J. Price, W. G. Smyth and S. A. Hewitt, Residues of Veterinary Drugs in Food (Part1), Proceedings of EuroResidue VI, 2008, 99.
7. Commission Decision 2002/657/EC, implementing Council Directive 96/23/EC concerning the performance of analytical methods and the interpretation of results, *Official Journal of the European Communities*, L221/8, August 17, 2002.
8. U. Flenker and A. Cawley, *J. Mass Spectrom.*, 2008, **43**, 854.
9. M. Hebestreit, U. Flenker, C. Buisson, F. André, B. Le Bizec, H. Fry, M. Lang, A. Preiss Weigert, K. Heinrich, S. Hird and W. Schänzer, *J. Agric. Food Chem.*, 2006, **54**, 2850.
10. CRL guidance paper (7 December 2007), CRLs view on state of the art analytical methods for national residue control plans.
11. http://www.vet-residues-committee.gov.uk/Papers/Papers08/VRC0841.pdf, accessed 24 March 2009.
12. http://www.vet-residues-committee.gov.uk/Reports/vrcar2006.pdf, accessed 24 March 2009.
13. USDA/FSIS, Screening for Zeranol by Elisa, Method CLG-ZRL.02, *Chemistry Laboratory Guidebook*, United States Department of Agriculture, Food Safety & Inspection Service, http://www.fsis.usda.gov/PDF/CLG_ZRL_02.pdf, 2007, accessed July 15 2008.
14. USDA/FSIS, Screen for Melengestrol Acetate in Bovine Fat using ELISA, Method CLG-MGA2.01, *Chemistry Laboratory Guidebook*, United States Department of Agriculture, Food Safety & Inspection Service, 2006.
15. USDA/FSIS, Confirmation of Melengestrol Acetate in by APCI/LC/MS, Method CLG-MGA1.00, *Chemistry Laboratory Guidebook*, United States Department of Agriculture, Food Safety & Inspection Service, 2003.
16. http://www.biacore.com/food/food_analysis/livestock/drug_residues/overview/index.html, accessed 24 March 2009.
17. http://www.biocop.org/content_pdf/WP8_Brochure_Final.pdf, accessed 24 March 2009.
18. T. F. H. Bovee, H. H. Heskamp, A. R. M. Hamers, R. L. A. P. Hoogenboom and M. W. F. Nielen, *Anal. Chim. Acta*, 2005, **529**, 57.
19. M. W. F. Nielen, E. O. van Bennekom, H. H. Heskamp, J. A. van Rhijn, T. F. H. Bovee and L. A. P. Hoogenboom, *Anal. Chem.*, 2004, **76**, 6600.
20. T. F. H. Bovee, G. Bor, H. H. Heskamp, R. L. A. P. Hoogenboom and M. W. F. Nielen, *Food Addit. Contam.*, 2006, **23**, 556.
21. M. W. F. Nielen, T. F. H. Bovee, H. H. Heskamp, J. J. P. Lasaroms, M. B. Sanders, J. A. van Rhijn, M. J. Groot and L. A. P. Hoogenboom, *Food Addit. Contam.*, 2006, **23**, 1123.
22. T. F. H. Bovee, J. P .M. Lommerse, A. A. C. M. Peijnenburg, E. A. Fernandes and M. W. F. Nielen, *J. Steroid Biochem. Mol. Biol.*, 2008, **108**, 121.

23. T. F. Bovee, R. J. Helsdingen, A. R. Hamers, M. B. van Duursen, M. W. Neilen and R. L. Hoogenboom, *Anal. Bioanal. Chem.*, 2007, **389**, 1549.
24. J. C. Rijk, T. F. Bovee, M. J. Groot, A. A Peijnenburg and M. W. Nielen, *Anal. Bioanal. Chem.*, 2008, **392**, 417.
25. U.S. Federal Register, 22 Feb. 1977, 42, No. 35.
26. J. A. Sphon, *J. Assoc. Offic. Anal. Chem.*, 1978, **61**, 1247.
27. A. L. Donoho, W. S. Johnson, R. F. Sieck and W. L. Sullivan, *J. Assoc. Offic. Anal. Chem.*, 1973, **56**, 785.
28. J. J Ryan and J. C. Pilon, *J. Assoc. Offic. Anal. Chem.*, 1976, **59**, 817.
29. K. A. Kohrman and J. MacGee, *J. Assoc. Offic. Anal. Chem.*, 1977, **60**, 5.
30. R. Baldwin, R. A. Bethem, R. K. Boyd, W. A. Buddle, T. Cairns, R. D. Gibbons, J. D. Henion, M. A. Kaiser, D. L. Lewis, J. E. Matusik and J. A. Sphon, *J. Am. Soc. Mass Spectrom.*, 1997, **8**, 1180.
31. R. Bethem, J. Boison, J. Gale, D. Heller, S. Lehotay, J. Loo, S. Musser, P. Price and S. Stein, *J. Am. Soc. Mass Spectrom.*, 2003, **14**, 528.
32. F. André, K. K. G. Wasch, H. F. DeBrabender, S. R. Impens and L. A. M. Stolker, *TrAC Trends Anal. Chem.*, 2001, **20**, 435.
33. E. J. H. M. Jansen, F. Freudenthal, H. J. van Rossum, J. L. M. Litjens and R. W. Stephany, *Biomed. Environ. Mass Spectrom.*, 1986, **13**, 245.
34. C. H. van Peteghem, M. F. Lefevere, G. M. de Haver and A. P. de Leenheer, *J. Agric. Food Chem.*, 1987, **35**, 228.
35. R. Bagnati, M. G. Castelli, L. Airoldi, M. P. Oriundi, A. Ubaldi and R. Fanelli, *J. Chromatogr.*, 1990, **527**, 267.
36. H. J. van Rossum, M. H. Blokland, S. S. Sterk and L. A. van Ginkel, Proceedings EuroResidue VI, The Netherlands, 2008, 627.
37. L. F. Krzeminski, S. Geng and B. L. Cox, *J. Assoc. Offic. Anal. Chem.*, 1976, **59**, 507.
38. AOAC International, Official Methods of Analysis, 17th edn. (on-line version), http://www.eoma.aoac.org/, 2008, accessed July 21 2008.
39. D. Arnold and R. Stephany, Residues of some veterinary drugs in animals and foods, FAO Food & Nutrition Paper, 2000, **41/13**, 75.
40. USDA/FSIS, Determination of Melengestrol Acetate in Bovine Fat, Method CLG-MGA.03, *Chemistry Laboratory Guidebook*, United States Department of Agriculture, Food Safety & Inspection Service, http://www.fsis.usda.gov/PDF/CLG_MGA_03.pdf, 2008, accessed 24 March 2009.
41. M. T. Andresen and A. C. Fesser, *J. AOAC Int.*, 1996, **79**, 1037.
42. CFIA, Private communication, 2008.
43. USDA/FSIS, *Chemistry Laboratory Guidebook*, United States Department of Agriculture, Food Safety & Inspection Service, http://www.fsis.usda.gov/Science/Chemistry_Lab_Guidebook/index.asp, 2008, accessed 24 March 2009.
44. T. M. Chichila, P. O. Edlund, J. D. Henion and R. L Epstein, *J. Chromatogr.*, 1989, **488**, 389.

45. O. W. Parks, R. J. Shadwell, A. R. Lightfield and R. J. Maxwell, *J. Chromatogr. Sci.*, 1996, **34**, 353.
46. A. A. M. Stolker, P. W. Zoontjes, P. L. W. J. Schillens, P. R. Kootstra, L. A. van Ginkel, R. W. Stephany and U. A. Th. Brinkman, *Analyst*, 2002, **127**, 748.
47. L. Giannetti, D. Barchi, F. Fiorucci, E. Gennuso, P. Sanna, M. Pallagrosi and B. Neri, *J. Chromatogr. Sci.*, 2005, **43**, 333.
48. S. -S. Hsu, T. R. Covey and J. D. Henion, *J. Liq. Chromatogr.*, 1987, **10**, 3033.
49. S. -S. Hsu, R. H. Eckerlin and J. D. Henion, *J. Chromatogr.*, 1988, **424**, 219.
50. E. H. Jansen, H. von Blitterswijk and R. W Stephany, *Vet. Quart.*, 1987, **6**, 60.
51. G. Moretti, G. Cavina, R. Alimenti, P. Cammarata and L. Valvor, *Anal. Chem. Symp. Ser.*, 1985, **23**, 471.
52. M. Rapp and H. H. D Meyer, *Arch. Lebensmittelhyg.*, 1987, **38**, 35.
53. E. A. I. Daeseleire, A. De Guesquière and C. H. Van Peteghem, *J. Chromatogr. Sci.*, 1992, **30**, 409.
54. R. W. Stephany, *APMIS*, 2001, **109**, S357.
55. E. Benoit, F. Garnier, D. Courtot and P. Delatour, *Ann. Rech. Vét.*, 1985, **16**, 379.
56. P. Silberzahn, L. Dehennin, I. Zwain and A. Reiffsteck, *Endocrinology*, 1985, **117**, 2176.
57. G. Maghuin-Rogister, A. Bosseloire, P. Gaspar, C. Dasnois and G. Pelzer, *Ann. Méd. Vét.*, 1988, **132**, 437.
58. G. Debruyckere, C. Van Peteghem, H. F. De Brabander and M. Debackere, *Vet. Q.*, 1990, **12**, 246.
59. G. Debruyckere and C. Van Peteghem, *J. Chromatogr.*, 1991, **564**, 393.
60. M. Vandenbroeck, G. Van Vyncht, P. Gaspar, C. Dasnois, P. Delahaut, G. Pelzer, J. De Graeve and G. Maghuin-Rogister, *J. Chromatogr.*, 1991, **564**, 405.
61. A. F. Rizzo, E. Alitupa, T. Hirvi, S. Berg, J. Hirn and A. Leinonen, *Anal. Chim. Acta*, 1993, **275**, 135.
62. J. D. MacNeil, J. A. Reid, C. D. Neiser and A. C. E. Fesser, *J. AOAC Int.*, 2003, **86**, 916.
63. L. A. van Ginkel, H. van Blitterswijk, P. W. Zoontjes, D. avn den Bosch and R. W. Stephany, *J. Chromatogr.*, 1988, **445**, 385.
64. G. W. Stubbings, A. D. Cooper, M. J. Shepherd, J. M. Croucher, D. Airs, W. H. H. H. Farrington and G. Shearer, *Food Addit. Contam.*, 1988, **15**, 293.
65. R. Bagnati and R. Fanell, *J. Chromatogr.*, 1991, **347**, 325.
66. M. Horie and H. Nakazawa, *J. Chromatogr.*, 2000, **882**, 53.
67. M. H. Blokland, P. W. Zoontjes, S. S. Sterk, R. W. Stephany, J. Zweigenbaum and L. A. van Ginkel, *Anal. Chim. Acta*, 2008, **618**, 86.
68. A. A. M. Stolker, S. H. M. A. Linders, L. A. vanGinkel and U. A. T. Brinkman, *Anal. Bioanal. Chem.*, 2004, **378**, 1313.

69. Evaluation of certain veterinary drug residues in food, Thirty-second Report of the Joint FAO/WHO Expert Committee on Food Additives, *Technical Report Series* **763**, World Health Organization, Geneva, 1988.
70. J. E. Roybal, R. K. Munns, W. J. Morris, J. A. Hurlbut and W. Shimoda, *J. Assoc. Off. Anal. Chem.*, 1988, **71**, 263.
71. T. R. Covey, D. Silvestre, M. K. Hoffman and J. D. Henion, *Biomed. Environ. Mass Spectrom.*, 1988, **15**, 45.
72. USDA, Determination and Confirmation of Diethylstilbestrol (DES) and Zeranol by GC/MS, *Chemistry Laboratory Guidebook*, United States Department of Agriculture, Food Safety & Inspection Service, http://www.fsis.usda.gov/PDF/CLG_ANA_02.pdf, 2008, accessed 24 March 2009.
73. CAC, Compendium of Methods of Analysis Identified as Suitable to Support Codex MRLs Developed by the Codex Committee on Residues of Veterinary Drugs in Foods, 2008.
74. L. C. Dickson, J. D. MacNeil, J. Reid and A. C. E. Fesser, *J. AOAC Int.*, 2003, **86**, 631.
75. USDA/FSIS, 2007 FSIS National Residue Sampling Program Sampling Plans. United States Department of Agriculture, Food Safety & Inspection Service, http://www.fsis.usda.gov/Science/2007_Blue_Book/index.asp, 2008, accessed 24 March 2009.
76. C. Akre, R. Fedeniuk and J. D. MacNeil, *Analyst*, 2004, **129**, 145.
77. R. W. Fedeniuk, M. West, R. Gedir, M. Mizuno, C. Neiser and J. D. MacNeil, *J. AOAC Int.*, 2006, **89**, 576.
78. R. M. Costain, A. C. E. Fesser, D. McKenzie, M. Mizuno and J. D. MacNeil, *Food Addit. Contam.*, 2008, **25**, 1520.
79. R. W. Fedeniuk, J. O. Boison and J. D. MacNeil, *J. Chromatogr. B*, 2004, **802**, 307.
80. C. J. M. Arts, M. J. Van Baak and J. M. P. Den Hartog, *J. Chromatogr. B*, 1991, **564**, 429.
81. FAO, Food and Nutrition Paper 41/12: Residues of Some Veterinary Drugs in Animals and Foods, Food & Agriculture Organization of the United Nations (FAO), Rome, Italy, 37–90, 2000.
82. S. J. Lehotay, A. DeKok, M. Hiemstra and P. Van Bodegraven, *J. AOAC Int.*, 2005, **88**, 595.
83. J. Chrusch, S. Lee, R. Fedeniuk and J. O. Boison, *Food Addit. Contam.*, 2008, **25**, 1482.
84. Council Directive 96/23/EC of 29 April 1996 on measures to monitor certain substances and residues thereof in live animals and animal products and repealing Directives 85/358/EEC and 86/469/EEC and Decisions 89/187/EEC and 91/664/EEC, Official Journal of the European Communities, **L125/8**, May 23, 1996.
85. S. A. Hewitt, M. Kearney, J. W. Currie, P. B. Young and D. G. Kennedy, *Anal. Chim. Acta*, 2002, **473**, 99.
86. P. R. Kootstra, P. W. Zoontjes, E. F. van Tricht and S. S. Sterk, *Anal. Chim. Acta*, 2007, **586**, 82.

87. R. Galarini, A. Piersanti, S. Falasca, S. Salamida and L. Fioroni, *Anal. Chim. Acta*, 2007, **586**, 130.
88. C. Van Poucke, M. Van De Velde and C. Van Peteghem, *J. Mass Spectrom.*, 2005, **40**, 731.
89. P. W. Zoontjes, P. R. Koostra, E. F. van Tricht and S. S. Sterk, Proceedings EuroResidue VI, 2008, 661.
90. H. Noppe, B. Le Bizec, K. Verheyden and H. F. de Brabander, *Anal. Chim. Acta*, 2008, **611**, 1.
91. K. L. Wubs, M. H. Blokland, S. S. Sterk and L. A. van Ginkel, Proceedings EuroResidue VI, 2008, 621.
92. DAFF National Residue Survey, Private communication, 2008.
93. L. Stolker, E. Oosterink, P. Rutgers, J. Laseroms, R. Peters, H. Mol, H. van Rijn and M. Nielen, Proceedings Euroresidue VI, 2008, 717.
94. P. W. Zoontjes, M. H. Blokland, S. S. Sterk, J. A. Zweigenbaum and L. A. van Ginkel, Proceedings EuroResidue VI, 2008, 633.
95. D. G. Kennedy, S. A. Hewitt, J. D. G. McEvoy, J. W. Currie, A. Cannavan, W. J. Blanchflower and C. T. Elliot, *Food Addit. Contam.*, 1998, **15**, 393.
96. K. M. Cooper, M. Tuomola, S. Lahdenpera, T. Lovgren, C. T. Elliott and D. G. Kennedy, *Food Addit. Contam.*, 2002, **19**, 1130.
97. K. M. Cooper, L. Ribeiro, P. Alves, V. Vozikis, S. Tsitsamis, G. Alfredsson, T. Lovgren, M. Tuomola, H. Takalo, A. Iitia, S. S. Sterk, M. Blokland and D. G. Kennedy, *Food Addit. Contam.*, 2003, **20**, 804.
98. F. M. Launay, L. Ribeiro, P. Alves, V. Vozikis, S. Tsisamis, G. Alfredsson, S. S. Sterk, M. Blokland, A. Iitia, T. Lovgren, M. Tuomola, A. Gordon and D. G. Kennedy, *Food Addit. Contam.*, 2004, **21**, 833.

CHAPTER 5

Current Research into New Analytical Procedures

Ed HOUGHTON,[a] PHIL TEALE,[a] EMMANUELLE BICHON[b] AND BRUNO LE BIZEC[b]

[a] HFL Sport Science, Newmarket Road, Fordham, Cambridgeshire, UK, CB7 5WW; [b] Laboratoire d'Etude des Résidus et Contaminants dans les Aliments (LABERCA), Ecole Nationale Vétérinaire de Nantes, Route de Gachet, BP 50707, 44307 NANTES Cedex 3, France

5.1 Introduction

Steroid hormones, including synthetic anabolic steroids and glucocorticoids, have been illegally used as growth promoters in livestock for a number of years,[1,2] although their use has been forbidden in the European Union (EU) since 1988 (88/146/EEC). Effective enforcement of the current EU prohibition of use of natural and synthetic steroids to promote growth in livestock requires development of efficient and cost-effective screening and confirmatory analysis techniques to support surveillance programmes.[3] The detection and confirmation of administration of natural steroid hormones to cattle and other species poses a particular significant analytical challenge and requires a detailed knowledge of both the endogenous steroid profiles in the biological fluids under study and the metabolism of the endogenous hormones of interest. It is relatively common, for example, for a parent drug to be rapidly converted to one or two major metabolites, in which case there is much to be said for targeting screening and confirmatory activities on the metabolites in question.

RSC Food Analysis Monographs No.8
Analyses for Hormonal Substances in Food-producing Animals
Edited by Jack F. Kay

Published by the Royal Society of Chemistry, www.rsc.org

Further, the relatively recent advent of illegal use of combined drug formulations exacerbates the difficulties of drug detection. In such circumstances, the concentrations of individual components may be significantly lower than those resulting from established treatments and thus may make demands on the limits of quantification of conventional screening methods beyond their capability. Other possible scenarios designed to circumvent current testing procedures include the use of designer drugs. For example, tetrahydrogestrinone (THG),[4] a synthetic steroid produced by the reduction of gestrinone, was deliberately designed to evade detection methods currently in use in sports testing laboratories and the possibility that similar approaches may be used to evade detection in the drug residue field cannot be ignored. This and the complexity in metabolism present the need to develop generic extraction and screening methods.

The abuse of THG in sport and the subsequent investigation of the use of this substance by elite athletes also highlighted the degree of sophistication employed to evade detection. Applying knowledge of the testing and sampling regime employed along with a detailed understanding of efficacy and pharmacology allowed a range of illegal substances to be used without detection. Given sufficient inducement, similar practices could be employed in meat production.

In addition, the innovative pharmaceutical and veterinary pharmaceutical industries continue to develop new and more potent drugs and, with the rapid advances in the biotechnology industry over the past decade, many of the drugs currently in development are protein based. The detection of protein-based drugs (*e.g.* recombinant growth hormone products and recombinant erythropoietin) is currently presenting serious challenges to conventional analytical approaches (*e.g.* chromatographic, hyphenated chromatographic/spectroscopic or direct immunochemical methods) in the field of drug control in human and animal sports.[5] Further challenges include the potential for stimulating endogenous production of growth-promoting protein and other target proteins through gene therapy.[6] For example, a group in the USA has developed and marketed, for research purposes, both a plasmid-based gene therapy and a delivery system for stimulating gonadotrophin releasing hormone production in the stallion.[7]

The veterinary drug residue field is also facing similar challenges. The potential abuse of potent human pharmaceutical preparations, the administration of cocktails and improvements in drug delivery systems all result in very low concentrations of drug residues and thus present challenges to conventional analytical approaches. These challenges will undoubtedly increase in the future with the potential for abuse of products from the biotechnology industry and registration of such products for veterinary use.

Thus there is a need to consider the application of recent developments in classical approaches to address these challenges or to consider alternative approaches to screening and confirmatory analysis. The classical approaches to control the misuse of drugs rely on the monitoring of the parent drug and its metabolites in urine, hair or tissue samples. Whereas gas chromatography tandem mass spectrometry (GC-MS or GC-MS/MS)[8,9] tends to be the technique for the unambiguous identification of anabolic steroids, particularly when screening on the basis of the presence of metabolites or a combination of the parent drug and metabolite, liquid chromatography (LC-MS/MS)[10,11] is preferred for glucocorticoids due to their polarity. However, API-LC-MS with multiple reaction monitoring is being increasingly applied in the field of veterinary drug residue analysis and provides a specific and sensitive approach to the detection of the parent anabolic steroids and some of their metabolites. The majority of the parent steroids studied have keto functions at C-3 and/or C-17, hydroxyl functions at C-17 and one or more sites of unsaturation. These unsaturated hydroxyl/keto steroids tend to show a good response under API-LC-MS conditions as indicated by the sensitivities of the developed methods but in cattle and other species many of these steroids are extensively metabolised resulting in little or no parent drug excreted in urine. For example, following administration of testosterone or nandrolone to cattle,[12,13] the major metabolites include saturated diols and isomers of 3,17-androstanediol and 3,17-oestranediol, which show a very poor response under LC-MS conditions. Thus in these cases, where little or no parent drug is excreted, LC-MS may not be the ideal technique for control of abuse and, hence, GC-MS is the favoured technique. Improvements in LC-MS technology, particularly the application of high-resolution MRM screening, in some cases in combination with ultra performance liquid chromatography (UPLC) have, however, provided significant improvements in specificity and sensitivity and provided very sensitive multi-residue methods for steroids in veterinary drug residue analysis. The approaches are discussed briefly in Section 5.2.

These increases in sensitivity in the LC-MS technology may not, however, be sufficient to detect administration of combined drug formulations nor can the approach be applied to designer drugs. MRM screening procedures, either low or high resolution, can only be targeted at lists of known steroids and not the unknown. Over the past few years, to address these challenges and those arising from the administration of protein based drugs, a significant amount of research has been devoted to investigating the use of biomarkers to detect drug administration both in veterinary drug residue analysis and in sport. A cell tissue, organ or biological fluid contains an ensemble of biomolecules which reflect normal body function. When the body is challenged by drug administration, this ensemble of biomolecules changes to reflect the functional activity of the administered drug. The application of principles based on monitoring the functional activity of the administered substance through the use of specific biomarkers thus present an attractive alternative to more conventional approaches to detect the drug or its metabolites. The biomarkers can be transcripts (mRNA), proteins or metabolites and the biomarkers are identified and monitored through the use of "omics" technologies: transcriptomics, proteomics and metabolomics. The application of molecular biology procedures has the potential to identify biomarkers in tissues to detect administration of illegal or unauthorised hormonal growth-promoting veterinary drugs and is discussed in Section 5.3

Xenosteroids (non-natural steroids) are nowadays efficiently monitored in cattle. However, the demonstration of misuse associated with the natural steroids testosterone, oestradiol or cortisol is much more problematic as no official concentration threshold or list of discriminative metabolites have ever been accepted and published by the European Commission (96/23/EC), making it almost impossible to follow the classical analytical approaches. In the absence of any concentration thresholds or discriminative metabolites, the analyst is faced with the challenge of developing methods to provide unequivocal discrimination between the natural presence of an endogenous hormone and its presence in the biological matrix arising from administration.

One of the unequivocal approaches currently used by some laboratories, whether in the field of antidoping[14–16] or food safety,[17] is to detect anabolic steroid administration by demonstrating the presence of the proprietary steroid esters at injection sites or in hair.[18] The detection of steroid ester in hair samples unambiguously demonstrates the administration of natural hormones and this approach along with measurement of the esters in injection sites, if any, is a direct strategy to confirm fraudulent practices.

For cortisol, there is no recognised unambiguous criterion to distinguish the exogenous from the endogenous metabolites; neither an official concentration threshold in bovine urine nor discriminative metabolites have been published and recognised officially by the competent authorities. The urinary concentrations of cortisol strongly depend on various factors including stress, sex, age, feeding, season and feedback control,[19] which complicates the setting of a reference threshold. However, a threshold value for the control of cortisol administration to horses has been determined[20] and accepted internationally.

The approach of the use for discriminative metabolites has been adopted to control boldenone administration in cattle. A meeting between experts of the Member States held in Brussels in 2003 decided that, on the basis of the scientific evidence available at the time, the presence of 17β-boldenone conjugates at any concentration in urine from veal calves was proof of illegal treatment.[21] Nowadays, the most promising approach for the control of natural steroid hormones is the $^{13}C/^{12}C$ ratio measurement of steroids by gas chromatography-combustion-isotope ratio mass spectrometry (GC-C-IRMS). In both human sports and in the control of veterinary drug residues in livestock, it has been shown that the administration of natural hormones (testosterone, oestradiol and cortisol) leads to an alteration of the $^{13}C/^{12}C$ ratio of their metabolites whereas the isotopic composition of precursor steroids (upstream in steroid metabolism pathway) remains unchanged. A significant difference in the isotopic composition between these so-called endogenous reference compounds (ERCs) and metabolites reveals an illegal administration. The few papers dealing with this approach in food-producing animals focus more on androgens, mainly testosterone,[22–26] and less on oestrogens[27] or cortisol.[28] The application of GC-C-IRMS for the control of these natural steroids is discussed in Section 5.4.

5.2 Some Applications of LC-MS Analysis

The commercial introduction of atmospheric pressure ionisation (API) in the mid 1980s was undoubtedly, at that time, the most significant development in LC-MS and provided the analyst with expectations of a truly robust, practical and sensitive LC-MS interface. These early expectations were justified and in the field of drug analysis the technique was rapidly accepted as the industry standard by the pharmaceutical industry in support of drug discovery and development. Now LC-MS impinges upon many scientific disciplines with a proven ability to handle compounds of diverse polarity and molecular weight.

Within the field of veterinary drug analysis, particularly for steroid hormones, acceptance of the technology was much slower than in the pharmaceutical industry possibly for two reasons: firstly there was the consideration of the poor response of the fully reduced metabolites of some of the steroid hormones under LC-MS conditions and secondly the comparative cost of the LC-MS systems when compared to bench-top GC-MS instruments.

However, over the past decade, there has been a marked increase in the use of LC-MS in the veterinary drug residue analysis field. This has been assisted by developments in API technology, developments in MS/MS, the introduction of lower-cost high-resolution instruments based on Time-of-Flight and Orbitrap technology and improvements in data acquisition and data processing in the form of multiple reaction monitoring and data-dependent scanning in specific time windows. These processes allow for screening of multiple analytes in a single LC run at high sensitivity and specificity.

The past, present and future of mass spectrometry in the field of veterinary drug residue analysis has been reviewed[29] and the use and advantages of LC-ToF-MS as a multi-residue screening method has been eminently covered in Chapter 4 of this book by Sharman *et al.* The authors have discussed the advantages of the medium to high resolution available from Time-of-Flight (ToF) technology and its application in the use of full scan techniques. The medium to high resolution available with ToF provides a marked increase in selectivity and hence sensitivity gains when compared to unit-resolution instruments.

Nielen *et al.*[30] have also discussed the advantages of applying accurate mass instrumentation to screening and confirmatory analysis in hormone and veterinary drug residue analysis. The authors compared the use of accurate mass ToF, Fourier transform ion cyclotron resonance and Fourier transform Orbitrap MS in a study for the mass resolution and accuracy requirements for LC/MS screening and confirmatory analysis of veterinary drug residues using stanozolol and clenbuterol as model compounds. For clenbuterol, the medium resolution accurate mass data obtained by ToF-MS/MS were confirmed by its analysis by FT Orbitrap MS. However, the greater resolving power of the LTQMS^2/FT Orbitrap (60,000 FWHM) was required for the LC-MS/MS^n analysis of stanozolol. The authors showed using FTMS that the majority of the product ions for stanozolol are doublets with only minor exact mass differences. These doublets are only partially resolved using Q-ToF-MS at 5000

(FWHM) mass resolution resulting in accurate but wrong average mass values for these ions, whereas using the LTQMS/FT Orbitrap these ions were resolved. An additional advantage of using accurate mass is that the elemental composition data generated can be used to identify unknown compounds. As a result of consideration of both theoretical and practical data, the authors have made a proposal for additional LC-MS criteria to be included in the Commission Decision 2002/657/EC for the use of accurate mass data in the analysis of veterinary drug residues.

Van der Heeft *et al.*[31,32] have also demonstrated the advantage of the higher resolving power of the Orbitrap MS in the analysis of steroid esters in bovine hair. Extracts of blank bovine hair were fortified with 14 steroid esters and the extracts were then analysed by UPLC-ToF-MS (resolving power approximately 10,000) and UPLC-Orbitrap MS (resolving power 7500 and 60,000). The resolution of 60,000 with a narrow mass tolerance window (5 ppm) was required to detect all the esters at the low $\mu g\,kg^{-1}$ concentration. When applying a resolution of 7500 on the Orbitrap or 10,000 on the ToF, accurate mass measurement was not sufficient to resolve analyte signals from those of isobaric compounds in the matrix resulting in errors of accurate mass measurement of >10 ppm.

Monteau *et al.*[33] have discussed the advantages of the application of LC-HRMS^n for growth-promoter control and demonstrated its ability for the identification of unknown compounds using clenbuterol as an example for structure elucidation and the screening and confirmatory analysis of steroid conjugates using boldenone sulfate as an example. The analysis of boldenone sulfate by LC-MS/MS and LC-HRMS^n has been discussed in more detail by Deceununck *et al.*[34]

Developments in the biotechnology industry over the past decade have resulted in the emergence of a number of recombinant protein-based drugs with the potential for abuse both in sport and in growth promotion in livestock. The development of direct methods to distinguish between the recombinant proteins and their natural homologues has presented the analyst with an interesting challenge, particularly when there is close homology between the recombinant and natural forms of the hormones. However, using high-resolution LC-MS/MS, Le Breton *et al.*[35] have developed a method to detect recombinant bovine somatotrophin (rbST) in bovine plasma. Bovine plasma (4 mL) fortified with rbST was diluted with phosphate buffer (0.1 M; pH 6.9) and proteins precipitated with 45% ammonium sulfate (15 hrs; 4 °C). The precipitate was resuspended in phosphate buffer and purified by SPE on a C4 column. The proteins of interest eluted from the column were then subjected to a second precipitation with methanol. The purified hormone was digested with trypsin and the digest analysed by high-resolution LC-MS/MS using the LTQ-Orbitrap at a resolution of 30,000. The MS was targeted at the N-terminal peptide fragment which contained amino acid sequence differences that allowed for the discrimination between rbST and the natural form of the hormone. Using this method, rbST was detected in plasma from a goat for up to 2 days after treatment with the recombinant protein.

Similar approaches have been used to detect the administration of growth hormone[36] and recombinant erythropoietin[37] to horses. The use of mass spectrometry to quantify proteins and protein biomarkers is also becoming more widely applied. Bobin *et al.*[38] used immunoaffinity methods to isolate IGF-I from horse plasma along with the internal standard, a recombinant IGF-I which differed slightly from natural IGF-I. The proteins were analysed while intact using ESI LC-MS to generate an envelope of multiply charged ions. These were then deconvoluted and the intensity of the combined ions used to quantify the analyte. The use of proteolytic peptides using heavy isotope labelled peptides as internal standards has also been extended to the quantitative analysis of proteins in blood plasma/serum.[39] Barton *et al.*[40] applied this approach to the identification of 72 of the most abundant proteins in equine plasma from which a method was developed to provide comparative quantification of 49 of these proteins using nanoflow LC.

While nanoflow LC-MS is a highly sensitive technique that provides the capability to analyse multiple peptides in a single analytical run, throughput is very limited. The application of UPLC to proteolytic peptide analysis promises the ability to handle a smaller set of analytes with very high throughput. Kay *et al.*[41] demonstrated that using this approach with sub 2 μm phases, multiple phospholipids could be reliably quantified in a single rapid run allowing 80 injections in less than 7 hours.

In the analysis of veterinary drug residues, matrix interference and co-eluting compounds can have a significant impact on the quality of the data produced, particularly when using HRMS where mass accuracy is very susceptible to these influences. The combination of ultra-performance liquid chromatography (UPLC) with HRMS can address this problem. UPLC, through the use of small particle (<2 μm) columns, provides significant improvement in resolution when compared to normal LC and also marked reductions in overall analysis times. Kaufmann *et al.*[42] have taken advantage of this combination in the development of a method for the quantitative analysis of over 100 veterinary drugs with a range of polarities in a variety of meat matrices (muscle, liver and kidney). The tissue samples were homogenised in two stages, acetonitrile followed by an aqueous extraction solution. A single stage reversed phase solid phase extraction with dedicated rinsing steps was used to provide purified extracts for analysis by UPLC-ToF. The mass spectrometer was operated at a resolution of 12,000 (FWHM) over the mass range 100–1000 m/z, the drugs being located by extracted exact mass chromatograms. An average of 50% of the drugs were detected below 1 $\mu g\,kg^{-1}$ and about 60% had analytical recoveries greater than 80%. The method was validated according to the Commission Decision 2002/657/EEC for the three tissue matrices. The authors highlighted some of the issues in the validation of such an assay with over 100 analytes and commented that there is a need to reconsider validation guidelines to address this type of multi-residue analysis.

Using UPLC in combination with a triple quadruple mass spectrometer, Abuin *et al.*[43] have developed a quantitative method for 6 thyreostats.

Purification of thyroid extracts was compared using SPE and gel permeation chromatography (GPC). Recovery, accuracy and precision were better using the GPC method, probably due to its amenability to automation and the overall method satisfied validation requirements of the Commission. The thyreostats were detected by multiple reaction monitoring using a single transition for each analyte.

Several other multi-residue methods have been reported recently using UPLC in combination with MS/MS or ToF-MS for a variety of combinations of matrices and drugs; basic and neutral pharmaceuticals in surface water;[44] pharmaceuticals in waste water;[45] tetracycline and quinolone antibiotics in pig tissue;[46] veterinary drugs in urine;[47] veterinary drugs in milk[48] and beta-agonists in bovine and porcine urine.[49] These publications indicate that the combination of UPLC with mass spectrometric techniques provides the analyst in the field of veterinary drug analysis with a powerful tool for the development of these multi-residue approaches.

5.3 Application of Biomarkers and "Omics" Technology to Detect Administration of Growth Promoters

In molecular biology, "omics" has become the suffix that denotes the study of the entire set of a class of biomolecules. Whereas in "classical" molecular biology, the unravelling of biological processes is studied by investigating the role of each component individually, the "omics" technologies can be used to study a biological process as a whole, by analysing all transcripts, proteins or metabolites in a cell, tissue or even organism. The study of the entire set of mRNA transcripts (transcriptomics), proteins (proteomics) and metabolites (metabolomics) has seen an enormous development in recent years. In human testing the abuse of gene therapy (known as gene doping) is perceived as a major future threat to the integrity of sports. The potential of the "omic" technologies to assist in the detection of gene doping has been recognised and a number of projects have been funded by the World Anti Doping Agency and potential detection strategies have been reviewed Baoutina *et al.*[6] The use of omics technologies to identify biomarkers of anabolic agents has also recently been reviewed by Riedmaier *et al.*[50]

The most established omics technology is transcriptomics. The successful human genome project[51] has resulted in the identification of tens of thousands of (in some cases putative) mRNA transcripts that are encoded by the genome. This information is used to generate DNA chips, which in turn are used to analyse the gene expression pattern of an organism/cell type/tissue of interest. Similar "gene chips" are available for a range of species including the bovine. Transcriptomics has the advantage that even genes that are expressed at low concentrations, but which may yet be very relevant to biological processes, may be detected. A disadvantage can be that the mRNA expression concentrations may not always accurately reflect the protein expression concentrations, due to post-transcriptional events (translation, post-translational modifications of proteins) that are not taken into account in transcriptomics.

Proteomics remains a major technological challenge. The number of different proteins is far greater than the number of transcripts that encodes these proteins. The most important reason for this is that post-translational modifications of proteins, such as phosphorylation, glycosylation or protein splicing, can result in the functional expression of several different proteins from a single transcript. Roughly estimated, each eukaryotic cell expresses approximately 100,000 different proteins. A further, and arguably more challenging, difficulty is the range of protein concentrations encountered in biological systems. For example, in plasma the range of protein concentrations covers approximately 12 orders of magnitude of which albumin represents approximately 50% of the total circulating protein while some of the signalling proteins are present at sub-picomolar concentrations. The oldest proteomics technology is 2D gel electrophoresis, developed more than two decades ago. Using this technique approximately 5000 of the most abundant proteins present in a protein extract can be detected. Using sample pre-fractionation procedures, this number can be increased. However, the low-abundance proteins are very difficult to detect. Alternative techniques that have been developed in more recent years include LC-MS applications for peptide analysis. Methods have been developed that allow the detection of proteins containing a specific feature, such as a phosphoryl group. However, even with these new proteomics methods, the sensitivity is still an issue.

Metabolomics is the study of all metabolites in a cell/tissue/organism. The number of different metabolites is probably even higher than the number of different proteins. The usual practice to analyse metabolites is to develop methods that detect a particular class of metabolites, such as organic acids, eicosanoids, lipids, and so on. Each method may detect hundreds to a few thousand of different molecules. In contrast to proteomics, the metabolomics methods that are applied are much more targeted towards the type of metabolites that are expected to be relevant to the scientific problem that is investigated.

In considering this challenge to conventional analytical approaches, much recent research has moved away from directly monitoring the drug and/or its metabolites to monitoring the output of its functionality at a molecular stage, *i.e.* a biomarker.

Proteomics, transcriptomics and metabolomics are well established but their application to the detection of administration of illegal substances or veterinary drugs to food-producing animals is a relatively new concept. The viability of the application of molecular biology techniques to veterinary drug residue analysis was first demonstrated at the Fourth International Symposium on Hormone and Veterinary Drug Residue Analysis.[52,53]

5.3.1 Transcriptomics

Transcriptomic analysis has been widely applied to increasing our understanding of the cells' response to growth promoters. Transcriptomic analysis typically takes one of two courses, either directly investigating candidate transcripts using real time PCR or similar targeted approaches or using gene arrays to provide more general coverage. Much of this work has been carried out in the commonly used laboratory animals such as rat and mouse, but a significant amount of research has directly addressed meat-producing species. For example Kamanga-Sollo *et al.*[54] attempted to shed light on the mechanisms by which anabolic steroids enhance muscle growth through measuring IGF-1 m-RNA levels in bovine satellite cells following exposure to 17β-oestradiol or trenbolone. Subsequently using real-time PCR analysis of muscle biopsy the time course of changes in muscle IGF-I, IGFBP-3, myostatin and hepatocyte growth factor (HGF) mRNA was studied by Pampusch.[55] IGF-I mRNA levels increased in treated animals over 26 days compared to control animals and initial samples from the treated individual. No change in the IGFBP-3, myostatin or HGF mRNA was observed. In an extension of this work Pampusch *et al.*[56] studied the effects not only on IGF-I, but IGF-I receptor, oestrogen receptor-alpha and androgen receptor mRNA levels in muscle. They concluded that oestradiol was responsible for the increased IGF-I mRNA level observed in steers implanted with a combined implant. Other examples are the investigation of the effect of growth hormone on hepatic expression of GH receptor m-RNA in the bovine[57] and oestrogenic receptor m-RNA expression in various bovine tissues following exposure to zeranol.[58] While these investigations improve our knowledge of the response of cells to growth promoters, and may in the long run help identify new biomarker targets, they do not directly address approaches to detection of abuse.

Other investigators have used similar approaches with more direct application to residue analysis. Reiter *et al.*[59] investigated changes in transcription levels in various bovine tissues (uterus, liver and muscle) using a candidate gene approach. In total 57 genes were investigated using RT-PCR following treatment with anabolic agents. Significant changes were identified in all tissues. In liver 17 of the 24 genes tested showed changes in regulation and in muscle up to 11 out of 17 depending upon the site and steroid tested. The uterus showed the largest changes in expression: 13 of the 29 genes tested were affected. The authors noted that further studies were required to take into account different animal husbandry and other affects but considered that the study was a preparatory step to developing a screening method based upon gene expression.

Pfaffl *et al.*[60] used rtPCR to investigate the effect of the synthetic progestagen melengestrol acetate on expression levels of steroid receptor, IGF-1 and its receptor mRNA in liver and muscle. A dose-dependent relationship between increasing melengestrol acetate concentration and mRNA expression was observed in liver for androgen and IGF-1 receptor and in neck muscularity for IGF-1.

Transcriptomic analysis using microarrays with multiple binding sites for each targeted m-RNA sequence was the first truly high-throughput approach applied to omic analysis of biological systems. Technological developments in so-called "next generation" or "massively parallel" sequencing[61] are likely to decrease the importance of this technique in the foreseeable future but for the time being transcriptomic microarray analysis is still widely applied in cellular chemistry research across many disciplines. The availability of appropriate microarrays is reliant upon the availability of an annotated genome and a market for the resultant gene chips. Arrays are currently available for a range of species including cattle, sheep, pigs, horses and chickens, although the extent of coverage varies considerably. Bendixen *et al.*[62] reviewed the potential application of this technology and highlighted the importance of experimental design, the choice of technologies and methods of analysis to be used.

Jeroen *et al.*[63] investigated the potential for using microarrays to detect changes in transcriptomic expression in liver cells from cattle treated with DHEA. Using unsupervised principal component analysis the profiles of the DHEA-treated animals were clearly distinct from those of the control animals with up to 579 genes differentially expressed, although this was reduced to 13 across all the experiments/control sets tested. The authors rightly pointed out the potential for biological variation such as age, climate, environment, *etc.* to confound transcriptomic approaches and highlighted the need to validate potential biomarkers in larger population studies.

Carraro *et al.*[64] reported the identification of a large number of differently expressed genes following administration of dexamethasone and oestradiol (18 male beef cattle 15–18 months old) at low concentrations to act as growth promoters. Of particular note was that one of the detected transcriptional changes, with an unknown function, resulted in a 67-fold up-regulation. It was also observed that the only down-regulated gene was that of myostatin, a protein that inhibits muscle growth.

While there is a relatively small number of publications directly related to the use of transcriptomic analysis targeted at the detection of growth-promoter abuse, interest in this approach is growing. While this is in part fuelled by interest and research in other areas of residue analysis/food safety, for example the BioCop project (www.biocop.org; European Commission contract FOOD-CT-2004-06988), the rapid development in genomic/transcriptomic tools is probably the primary force driving this. The power of the next-generation sequencing technologies[61] is set to transform the range and speed of genomic/transcriptomic analysis. It seems highly probable that as a result the early but as yet unmet promise of transcriptomic biomarkers will be realised in the foreseeable future.

5.3.2 Proteomics

The application of proteomics to biomarker discovery typically takes one of two courses: an informed approach where known physiological and biochemical knowledge is used to target potential biomarkers or an uninformed or "*de novo*" approach where broad spectrum analytical techniques are used to identify differences between treated and untreated individuals. Gardini *et al.*[65] applied two-dimensional electrophoresis to extracts of cytosols and microsomes from calves treated with growth-promoting agents. Adenosine kinase and reticulocalbin were found to be differentially expressed. The authors considered the results show the ability of the proteomic approach to find biomarkers of illicit growth promoter use with the potential to develop large-scale screening methods.

The importance of the growth hormone axis proteins and peptides as a potential indicator of the abuse of growth promoters has long been recognised and forms the basis of a proposed test for rhGH abuse in human plasma.[66] Renaville *et al.*[67] determined plasma levels of IGF-I, IGFBP-2, IGFBP-3 and thyroid hormones in dexamethasone-treated calves while Johnson *et al.*[68] analysed IGF-I and IGFBP concentrations following administration of a trenbolone acetate and oestradiol implant. Numerous other publications on the growth hormone axis in experimental and food-producing animals have been published; an in-depth review of these is beyond the scope of this publication. Renaville *et al.*[69] produced a review of the manner in which metabolism is regulated by the somatotropic axis using different examples including growth-promoter administration.

Following an oral dose of boldenone and boldione to veal calves, Draisci *et al.*[70] carried out proteomic analysis of plasma samples using two-dimensional electrophoresis and MALDI-ToF-MS LC-MS/MS procedures. Using a Western blot analysis an N-terminal truncated form of apolipoprotein A1 was confirmed as showing a time-dependent increase beyond the point at which the steroids were detectable in urine or plasma.

Mooney *et al.*[71] investigated perturbed profiles within a panel of biomarkers in blood from calves subjected to nortestosterone decanoate, 17β-oestradiol benzoate and dexamethasone administration. Markers studied included both proteins and metabolites, urea, aminoterminal propeptide of type III procollagen (a proposed biomarker of rhGH abuse in humans) and sex hormone binding globulin profiles were found altered in response to treatments. In a further refinement Mooney *et al.*[72] used two-dimensional gel electrophoresis to compare circulating protein profiles of treated and untreated calves. Following nandrolone/oestradiol administration alpha-2-antiplasmion precursor, serotransferrin precursor and endopin-1 were all found to be down-regulated and identified as potential biomarkers. Following administration of dexamethasone, serotransferrin precursor and fetuin-A were down-regulated and L-lactate dehydrogenase B and alpha-1-antitrypsin precursor were up-regulated. The authors concluded that the findings demonstrate the potential of using markers which cover a spectrum of biological activity.

Typically proteomic studies use well-controlled populations of treated and untreated animals for experimental purposes. In an interesting departure Biancotto *et al.*[73] investigated animals passing through abattoir for study. Following histological screening of target organs individual carcasses were assigned as negative, suspect or strongly suspect of being previously administered growth promoters. Muscle samples were selected from each group and analysed using two-dimensional electrophoresis. Suspect proteins were characterised using MALDI-ToF followed by proteolytic digestion and LC-MS/MS identification; 51 potential target proteins were identified.

Cacciatore *et al.*,[74] as part of the EU-funded Biocop project (www.biocop.org), studied the effect of the administration of growth-promoting agents including oestradiol, nandrolone and dexamethasone on nine candidate protein biomarkers in the bovine. Potential markers, immunoreactive inhibin (ir-inhibin), osteocalcin, insulin-like growth factor 1 (IGF-1), insulin-like growth factor-binding protein 2 (IGFBP-2), IGFBP-3, luteinising hormone (LH), follicle-stimulating hormone (FSH) and prolactin, were identified based upon previously published research and analysed using available immunoassays. A significant advantage of this targeted approach is that it allows low-concentration markers, unlikely to be detected using discovery proteomic approaches, to be quantified. Ir-inhibin and osteocalcin were identified as potential markers of androgen, oestrogen and glucocortico abuse. The authors considered that the use of a panel of markers, possibly analysed using a multiplexed approach, would provide greater sensitivity for abuse compared to individual markers. In order to provide a timely high-throughput approach to the analysis of multiple protein markers, van Meeuwen *et al.*[75] described a surface plasmon resonance biosensor capable of quantifying 16 analytes simultaneously with a run time of approximately 15 minutes.

A similar approach of targeting specific proteins to discover potential biomarkers of anabolic steroid misuse, in this case in the equine, was undertaken by Barton *et al.*[40] Due to the limited availability of immunoassays for equine proteins a multiplexed LC-MS assay was developed and applied. Clusterin and leucine-rich alpha-2-glycoprotein were found to increase over the course of testosterone administration. Many of the proteins described in the literature as putative biomarkers of steroids were not found to change, a situation also highlighted by Cacciatore *et al.*[74] This highlights the difficulty of comparing across species barriers, different treatment regimes, *etc.* The use of different immunoassays and protein standards is also a significant issue.

Clearly there is a large body of information relating to proteins and their relationship to nutrition, husbandry, use of growth promoters, *etc.* Much of this information has potential value for assisting researchers in identifying biomarkers of growth-promoter abuse. In addition, techniques for investigating novel protein biomarkers continue to develop. In human sports testing the biological variability of protein expression represents a hurdle to the introduction of protein biomarker-based screening and confirmatory methods. Typically, within breed, food-producing animals have lower genetic variability compared to humans and approaches to animal husbandry tend to be consistent within geographical regions. In addition, following slaughter, a range of tissue types is available for testing. Further, the increasing use of transcriptomic approaches is likely to provide new target proteins for investigation using proteomic techniques. Given these advantages the potential for the application of protein biomarkers to detection of growth-promoter abuse is expected to become a reality in the foreseeable future.

5.3.3 Metabolomics

Metabolomics is an emerging field of "omics" research that focuses on large-scale and high-throughput measurement of small molecules (metabolites) in biological matrices. The metabolome is the collection of all small-molecule metabolites or chemicals that can be found in a cell, organ or organism and as such is ideally positioned to be used in many areas of food science and nutrition research. Metabolomic research (also known as metabonomics or metabolic profiling) has only become possible as a result of recent technological breakthroughs in small molecule separation and identification. These include robust MS instruments suitable for precise mass determination and nuclear magnetic resonance (NMR) spectrometers. While NMR offers rapid, highly selective and non-destructive measurements, it also exhibits relatively low sensitivity and mass spectrometry measurement following chromatographic separation potentially offers the best combination of both sensitivity and selectivity. GC/EI-MS was one of the first methods involved in metabolic profiling[76] but other techniques such as metastable atom bombardment (MAB) ionisation mode are known to provide selective ionisation as well as controlled fragmentation processes. Until recently, most of the work in this area has focused primarily on clinical or pharmaceutical applications such as drug discovery, drug assessment, clinical toxicology and clinical chemistry. However, over the past few years, metabolomics has also emerged as a field of increasing interest to food and nutrition scientists.[77] Lommen *et al.*[78] described a data analysis strategy for identifying unknown compounds with potential application to contaminant analysis and went on to propose such an approach as applicable to a range of problems in the area of residue analysis,[79] although application to detection of growth-promoter abuse was not demonstrated.

Dumas *et al.*[80] used pyrolysis coupled to metastable atom bombardment (MAB) and Time-of-Flight mass spectrometry (Py-MAB-ToFMS) for the first time as a metabolomic tool, providing a suitable variable generator for assessing weak physiological variations induced by normal or subnormal anabolic treatment conditions in cattle. This work demonstrated how indirect metabolic variations induced by physiological responses to hormonal treatment can be evidenced through urine monitoring without searching for specific hormone residues. Furthermore, this paper was the first to demonstrate the suitability of metabolomic approaches as a powerful screening tool for anabolic steroid use in cattle with further potential applications in the monitoring of sport and horseracing control.

Antignac *et al.*[81] investigated the applicability of metabolomics using mass spectrometric approaches in the area of food safety including the use of anabolic steroids in cattle. They concluded that while the technique had promise and was potentially a powerful future screening technology, significant technological and methodological issues needed to be addressed. In a novel approach, Cunningham *et al.*[82] undertook the targeted metabolic profiling of cattle sera following administration of growth promoters. Utilising standard clinical chemistry parameters and analysis using support vector machines, treated individuals were identified with high sensitivity and specificity.

Kieken *et al.*[83] investigated the potential for using mass spectrometric based metabolomics to detect growth hormone administration to the horse. Following filtration and freeze drying of the urine samples the extracts were reconstituted in water and analysed using an Orbitrap™ system operating at 30,000 resolution. Twenty-four ions were identified as highly discriminating using PCA. Although the study was limited to a very restricted population the ability to discriminate treated and untreated animals was encouraging, although the authors pointed out the need for further validation.

While it is clear that interest in the application of metabolomics to residue analysis is receiving increasing interest the volume of peer-reviewed publications in the area is restricted. The potential impact of metabolomics is currently limited by the available technology and availability of suitable databases. However, it is predicted that this is an area of research with great potential for adoption by future hormone control programmes.

5.3.4 Relevance of $^{13}C/^{12}C$ Measurement for Steroid Abuse Control

5.3.4.1 Principle

According to Farquhar *et al.*,[84] variation in the $^{13}C/^{12}C$ ratio is the consequence of "isotope effects", often classified as being either kinetic or thermodynamic. One example of interest is the difference between the kinetic constants for the reaction of $^{12}CO_2$ and $^{13}CO_2$ with ribulose bisphosphate carboxylase-hydrogenase (Rubisco). Indeed, during photosynthesis, the assimilation of carbon dioxide by plants occurs *via* two principle forms of metabolism, the C_3 metabolism (Calvin cycle) and the C_4 metabolism (Hatch and Slack cycle). These two photosynthesis mechanisms present a different type of isotope fractionation. Products of C_4 plants have higher concentrations of ^{13}C than similar products of C_3 plants.[84] Synthesised steroids are normally made from *Dioscorea* spp. or soy[85] which are C_3 plants. Endogenously produced steroids derive from the diet, which is based on a C_3/C_4 mixture. Consequently, administered steroids and subsequently their metabolites are depleted regarding their $^{13}C/^{12}C$ ratios compared to steroids endogenously produced by the animal.

5.3.4.2 ^{13}C Measurement

The ^{13}C content is determined by monitoring the carbon dioxide produced following the complete combustion of the analyte of interest. The abundance of the principle isotopomers of masses 44, 45 and 46, resulting from the different possible combinations of isotopes ^{18}O, ^{17}O, ^{16}O, ^{13}C and ^{12}C, are determined from the ionic currents measured by three different collectors of a mass isotopic spectrometer. The detected current is proportional to the respective quantity collected in each of the detectors monitoring m/z 44, 45 and 46.

Consequently, the mass spectrometer must be perfectly calibrated to collect the accurate quantity of each isotopomer. Thus the m/z 44 collector collects the ions $^{12}C^{16}O^{16}O^{+\bullet}$ and m/z 45 collector collects the ions $^{13}C^{16}O^{16}O^{+\bullet}$ but also $^{12}C^{17}O^{16}O^{+\bullet}$ and $^{12}C^{16}O^{17}O^{+\bullet}$. For accurate determination of ^{13}C, measuring the abundance of these ions allows the isotope ratio $^{13}C/^{12}C$ to be calculated, after correcting the contribution of ^{17}O to the ion beam at mass 45.[86] The formula used is presented in Figure 5.1.

The m/z 46 corresponds to the sum of ions $^{12}C^{16}O^{18}O^{+\bullet}$, $^{12}C^{18}O^{16}O^{+\bullet}$, $^{13}C^{16}O^{17}O^{+\bullet}$, $^{13}C^{17}O^{16}O^{+\bullet}$ and $^{12}C^{17}O^{17}O^{+\bullet}$. The ionic species associating ^{13}C and ^{17}O are rare, so $^{12}C^{17}O^{17}O^{+\bullet}$, $^{13}C^{16}O^{17}O^{+\bullet}$ and $^{13}C^{17}O^{16}O^{+\bullet}$ isotopomers are not taken into account (the sum of the three corresponding to 0.224% of the $^{12}C^{16}O^{18}O^{+\bullet}$ isotopomer). In this way, the intensity recorded for m/z 46 corresponds to abundance of the main isotopomers $^{12}C^{16}O^{18}O^{+\bullet}$ and $^{12}C^{18}O^{16}O^{+\bullet}$ and allows for the measurement of ^{18}O contribution and to deduce ^{17}O contribution.

Comparison with a calibrated reference against the international reference, Vienna-Pee Dee Belemnite (V-PDB), allows for the calculation of ^{13}C content on the $\delta^{13}C$ (deviation in stable isotope composition) relative scale. V-PDB is defined as follows:[87] V-PDB is the primary reference material for measuring natural variations of ^{13}C isotope content, consisting of calcium carbonate from a Cretaceous belemnite guard from the Pee Dee Formation in South Carolina (USA). Its $^{13}C/^{12}C$ isotope ratio or (R_{PDB}) is 0.0112372. PDB reserves have been exhausted for a long time, but it has remained the primary reference for expressing natural variations of ^{13}C isotope content and against which the reference material available at the International Atomic Energy Agency (IAEA) in Vienna (Austria) is calibrated. Isotopic indications of naturally occurring ^{13}C are conventionally expressed in relation to V-PDB with the following formula (*cf.* Figure 5.2).

$$\frac{^{17}O}{^{16}O} = K\left(\frac{^{18}O}{^{16}O}\right)^{a} \qquad \begin{array}{c} a = 0{,}516 \\ K = 0{,}0099235 \end{array}$$

Figure 5.1 Craig correction formula.

$$\delta^{13}C(‰) = \frac{(R_{sample} - R_{standard})}{R_{standard}} \times 10^3$$

Figure 5.2 Formula for the determination of the deviation in stable isotope composition (R corresponds to abundance ratio $^{13}C/^{12}C$; $R_{standard} = 0.0112372$ for V-PDB reference, $\delta^{13}C_{V-PDB} = 0‰$).

5.3.4.3 Application to Steroid Control

The GC-C-IRMS strategy was developed for doping control in sport for the first time in 1994.[88] The isotopic composition of precursors of administered steroids as well as steroids on a different metabolic pathway (*e.g.* corticosteroids and androgen metabolic pathways) remains unchanged after administration and can be used as endogenous reference compounds (ERCs). If the difference between the $\delta^{13}C_{VPDB}$ values of a steroid or its metabolites and the ERC exceeds a given limit, this is considered as an evidence for the presence of exogenous steroids (*cf.* Figure 5.3, cases a and c). If this threshold is not

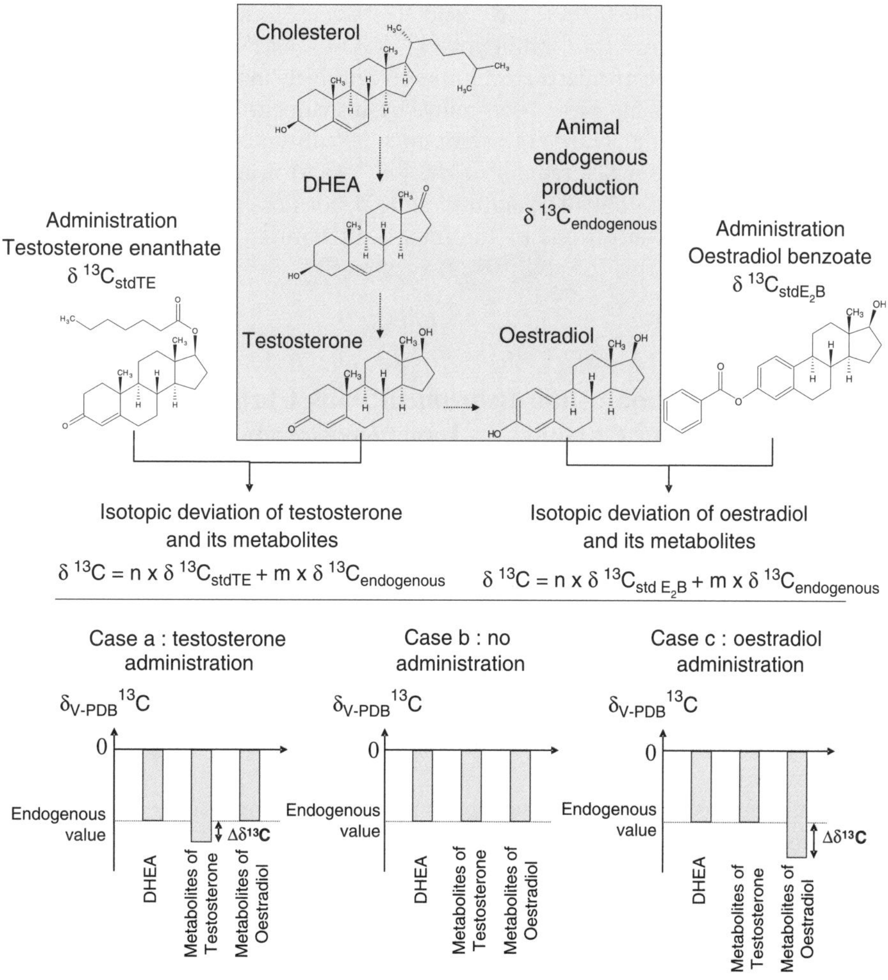

Figure 5.3 Variation of the isotope deviation value after administration of testosterone and oestradiol esters (n: quantity of molecules injected; m: quantity of molecules produced by the animal).

exceeded (case b), the difference between the ERC and metabolite is then attributed to a natural variability; the sample is then concluded compliant. In the anti-doping control in sport for anabolic and androgenic steroid abuse, a recommended threshold set at 3‰ is currently used.[89]

In parallel, a hyphenated approach based on steroid $^{13}C/^{12}C$ isotopic ratio measurement by GC-C-IRMS appears promising for cortisol abuse control. A significant difference in the isotopic composition between these so-called endogenous reference compounds (ERCs) and metabolites highlights the illegal administration of cortisol. The literature includes some papers dealing with cortisol control in athletes' urine.[90–92] A specific method dedicated to natural corticosteroids in cattle has been published by Bichon *et al.*[28]

In addition, GC-C-IRMS is currently used to determine the origin of numerous organic substances (fatty acid,[93,94] sugar,[95] ascorbic acid,[96] PAH,[97] POPs[98]) and to assure food authenticity.[99] GC-C-IRMS, in relation to other hyphenated mass spectrometric techniques is relatively insensitive in spite of the fact that improvements have been made by instrument providers and others over the past decade. Nanogram amounts of substances are mandatory to provide an accurate measurement of the $^{13}C/^{12}C$. However, the technology is compatible with the control of natural steroid hormones in cattle due to sufficient urinary concentrations of the important natural androgenic and anabolic steroids.

5.3.5 Presentation of the Instrument: Gas Chromatograph Coupled to Combustion Interface Isotope Ratio Mass Spectrometer (GC-C-IRMS)

5.3.5.1 The Gas Chromatograph, the Combustion Interface and the Mass Spectrometer

Splitless injection of purified samples is preferred to any other system; however, recent papers[100,101] have discussed the use of large volume injector to enhance the fraction of the injected sample thus improving the sensitivity. Non-polar stationary phase (100% methylpolysiloxane or 5% phenylpolysiloxane) are used (minimum 30 m × 0.25 mm×0.25 μm), generally with helium as carrier gas.

Compounds eluting from the chromatographic column pass through the combustion furnace reactor, an alumina tube containing Cu, Ni and Pt wires maintained at least at 850 °C, where they are oxidised producing CO_2 and H_2O. The water is removed either by passing the gases through a liquid nitrogen water removal trap (at –100 °C with a thermocouple) or by a Nafion® membrane which is permissive only to water. The remaining CO_2 passes into the isotope ratio mass spectrometer, an electron ionisation source operating at 100 eV. Ions (m/z 44, 45 and 46) are separated using a magnetic sector and detected by three specific Faraday collectors.

5.3.5.2 Calibration of the GC-C-IRMS Instrument

For good precision, isotope ratio measurement requires a stable, repeatable and linear response of the instrument. These parameters can be affected by water (brake of the cold trap) or air intake in the GC-C-IRMS, or by a fouling of the system. To guarantee the performances of the instrument, daily monitoring is recommended to check all the instrumental parts. The isotope ratio mass spectrometer is controlled with a reference gas which is directly introduced in the source. The overall GC-C-IRMS system, and its operation, must also be controlled with the use of steroid reference solutions throughout the analytical process. It is important to provide daily checks for stability, linearity and the overall system to guarantee the accuracy of GC-C-IRMS measurement.

5.3.5.2.1 Stability. The isotope ratio mass spectrometer (Figure 5.4) must be equipped with a dual inlet, to measure alternately the unknown analyte and the CO_2 reference gas. The CO_2 reference gas, for which the $^{13}C/^{12}C$ isotope ratio is known, is injected by "pulses".

To check the stability of the instrument, ten pulses are released consecutively until the variability between the ten consecutive isotopic deviation values is < 0.5‰. The CO_2 pressure is set to reach an intensity value corresponding to the middle of the linearity range, *i.e.* around 7 nA (corresponding to a pressure of 8 psi). Figure 5.5 presents an acceptable result of stability with a maximal deviation of 0.2‰ between the second and the ninth isotope ratio values. Finally, to check the stability during each sample run, CO_2 pulses are then introduced in each unknown sample acquisition (three pulses at the beginning of each run) to control the stability of the instrument during sample analysis.

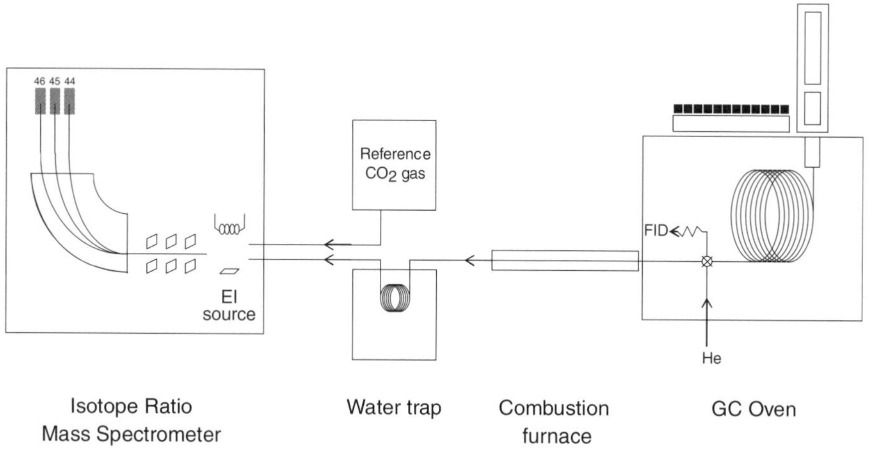

Figure 5.4 Illustration of an isotope ratio mass spectrometer coupled to a gas chromatograph *via* a combustion interface.

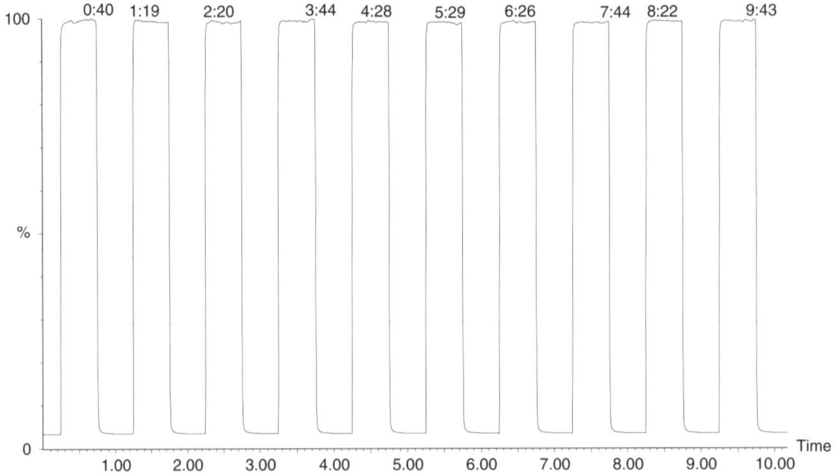

Peak No	Major Height (nA)	RT (Sec)	Ratio 45/44
1	7.69	30.0	1.1864E-02
2	7.69	90.0	1.1864E-02
3	7.64	150.0	1.1863E-02
4	7.69	209.9	1.1863E-02
5	7.68	270.0	1.1863E-02
6	7.67	330.0	1.1863E-02
7	7.64	390.0	1.1863E-02
8	7.65	450.0	1.1863E-02
9	7.68	510.0	1.1862E-02
10	7.67	569.9	1.1862E-02

Figure 5.5 Example of stability results with 10 CO_2 pulses.

5.3.5.2.2 Linearity. To check the linearity of the mass spectrometer, five pulses of the reference gas are released consecutively until the variability between the five consecutive isotopic deviation values is < 0.8‰. The CO_2 pressure is changed five times in all the linearity range of the instrument, *i.e.* between 2 and 12 nA in this example (corresponding to a range pressure of 4–15 psi). Figure 5.6 presents an acceptable result of linearity with a maximal deviation of 0.3‰ between the first and the fifth isotope ratio values.

5.3.5.2.3 The Overall Instrument Performance – Control Chart on Certified Steroids. Samples were measured against the laboratory's working standard (DHEA acetate, testosterone acetate, 5-androstene-3β,17β-diol diacetate, with their certified $\delta^{13}C_{V-PDB}$ values). These compounds are injected on the GC-C-IRMS system before, during and after each daily sequence of injections to control the measurement precision of the isotope ratio for all the samples. Figure 5.7 presents the different critical points controlled during this step. At first, the peak resolution is assessed. Symmetry and separation of chromatogram peaks have to be sufficient to guarantee the isotope ratio measurement of each compound separately. Secondly, CO_2 pulses must

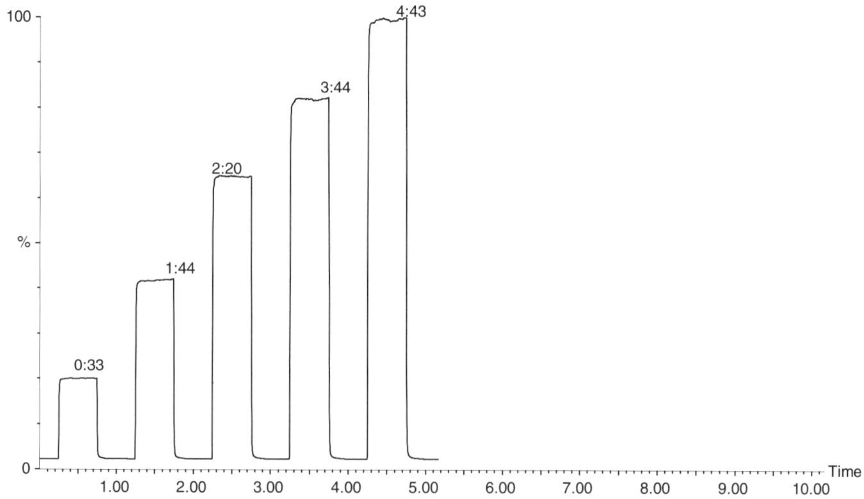

Peak No	Major Height (nA)	RT (Sec)	Ratio 45/44
1	2.09	30.1	1.1853E-02
2	4.67	90.0	1.1855E-02
3	7.42	150.0	1.1855E-02
4	9.44	209.9	1.1855E-02
5	11.53	269.9	1.1856E-02

Figure 5.6 Example of linearity results with 5 pulses.

demonstrate that the IRMS response is stable (as reported in Section 5.3.5.2.1). Finally, each isotopic deviation is controlled against a certified compound (here testosterone acetate) in reporting the isotopic deviation on a control chart. The thin line corresponds to the alert values ($\pm 2 \times$ standard deviation) and the bold ones the critical values ($\pm 3 \times$ standard deviation). When the alert value is crossed, maintenance is envisaged on the instrument before any other injection. Clearly, the non-co-elution of target analytes is mandatory to provide an accurate measurement.

5.3.6 Sample Preparation Steps

5.3.6.1 Objectives

Sample preparation includes in most papers[24–27] hydrolysis of glucuronic acid conjugates followed by octadecyl-functionalised silica solid-phase extraction (C18 SPE), a liquid/liquid extraction, a solvolysis of sulfated steroids and an additional silica SPE (SiOH SPE) purification. The extract is then purified with a two-stage semi-preparative HPLC (dimethylaminopropyl followed by C18 functionalised

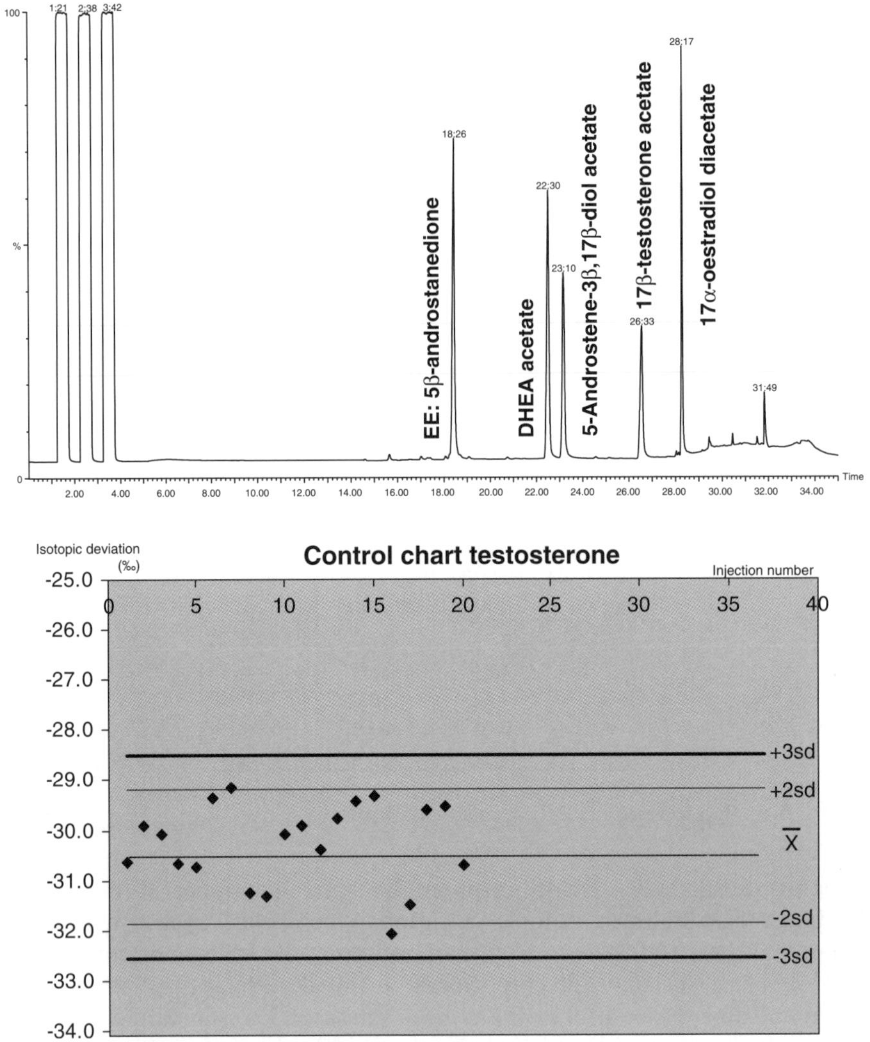

Figure 5.7 Typical GC separation of reference steroids (TIC). Control chart used and applied daily for each of the steroids (mean value −30.5‰, 2sd [−29.2‰; −31.9‰], 3sd [−28.5‰; −32.6‰]).

silica phases) and the purified steroids are then derivatised (acetylation) prior to GC-MS identification and GC-C-IRMS measurement (*cf.* Figure 5.8).

For glucocorticoids (metabolites of cortisol) in urine samples, published strategy includes reverse phase (copolymer styrenedivinylbenzene and N-vinylpyrrolidone) and silica SPE and an oxidation step followed by a GC-C-IRMS measurement.

The main objective of the sample preparation is to introduce into the IRMS analyser the analyte of interest with the highest purity, whilst minimising any

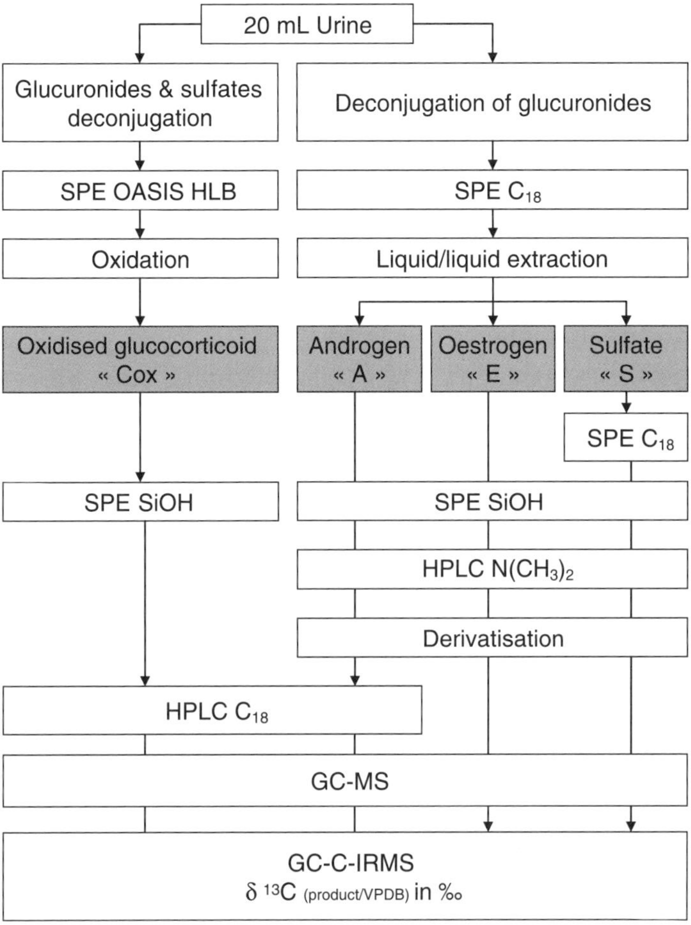

Figure 5.8 Analytical strategy for natural steroids and glucocorticoids abuse control.

isotopic fractionation of the analyte in the preparation process, thus maintaining its isotopic fingerprint. It is clearly important to minimise any co-elution of target analytes. Such co-elutions can be caused by matrix interferences which have been co-extracted, by contaminants coming from the process (phthalates, stationary phase) or target compounds with very similar chromatographic behaviour such as DHEA and epi-androsterone or 5-androstene-3β,17α-diol and 5α-androstane-3β,17α-diol. The second critical point is to minimise any isotopic fractionation which could induce an artificial depletion or enrichment in ^{13}C for the considered analyte and lead to a mistaken value. This phenomenon appears particularly during a chromatographic separation (solid-phase extraction, semi-preparative HPLC or GC),[27,102] the enriched ^{13}C isotopomers not possessing exactly the same physicochemical properties as ^{12}C corresponding ones.

These two critical points are generally taken into account during the development of the analytical method and controlled during the analytical process.

5.3.6.2 Hydrolysis of Steroid Conjugates

The enzymatic and chemical hydrolysis steps of sulfo- and glucuro-conjugates of androgens and oestrogens have been widely discussed in the literature.[24,103–105] The most critical steroid conjugates to hydrolyse enzymatically are well known to be the sulfo-conjugates. In this study, special attention has been paid to DHEA, which is known to be mainly sulfo-conjugated;[106] the other key target steroids are mainly glucurono-conjugates.

The different strategies of deconjugation, either chemical or enzymatic (with different sources of enzyme, *e.g. Helix pomatia* or *E. coli*), were assessed for their hydrolysis efficiency. *E. coli* was found to be the most efficient and specific way to deconjugate steroid glucuroconjugates and chemical solvolysis the most appropriate way to hydrolyse DHEA–SO_3H.[26] From a strategic point of view, the chemical hydrolysis was performed after two SPE and two LLE steps to minimise the production of interfering compounds. During the second LLE step, the oestrogens were separated from the aqueous layer so that only sulfate compounds underwent the chemical hydrolysis. In fact, the solvolysis is not suitable for oestrogens because of degradation of oestradiol during this process (about 30%).

Glucocorticosteroids in urine are conjugated with both glucuronic acid and sulfate. As current IRMS instruments have a limited sensitivity for steroids, hydrolysis of both types of conjugate is necessary in order to obtain the maximum quantity of free cortisol metabolites in the purified extract for GC-IRMS analysis. This was achieved using a β-glucuronidase aryl-sulfatase enzymatic mixture.[28]

5.3.7 Sample Purification

Because of their radical difference in terms of chemical behaviour, oestrogens and androgens were fractionated between sodium hydroxide and pentane at pH 14.[28] At this pH, phenolic steroids are converted into their phenolate form, their pKa value (oestradiol and oestrone) being 10.7 ± 0.1.[107] Phenolates were neutralised by addition of acetate buffer (pH 5.2), for further extraction into organic solvents.

For glucocorticoids purification, the stationary phase OASIS HLB was chosen.[44] Glucocorticoids can be preferentially eluted from the cartridge whereas oestrogens and androgens remain adsorbed on the SPE. The process provides an efficient sample preparation step.[28]

The purification by semi-preparative HPLC is an essential step before any GC-C-IRMS measurement of steroids at the low $ng\,ml^{-1}$ concentration in bovine urine samples. The need for such a strategy has been already reported in several articles.[103,104,107,108] A system involving a 3-dimethylaminopropyl-functionalised silica column was tested;[27] demonstration of the purification efficiency is shown on Figure 5.9. This step clearly allowed the powerful

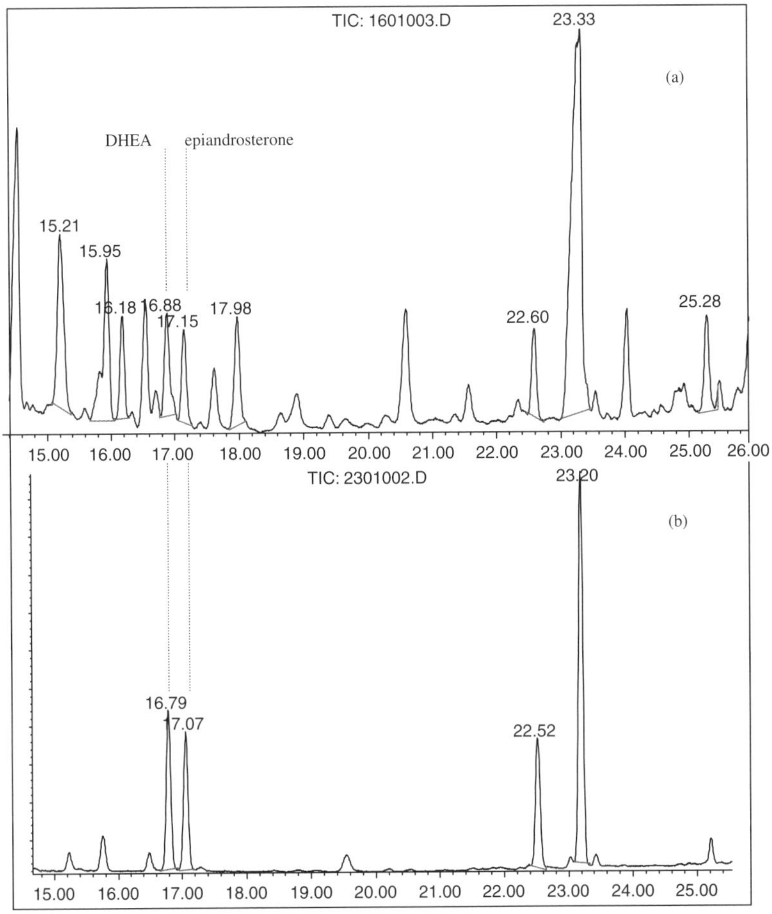

Figure 5.9 GC-MS chromatograms (TIC) of incurred DHEA and epi-androsterone in bovine urine observed without (a) and with (b) dimethylaminopropyl-functionalised silica semi-preparative HPLC.

clean-up of the wide range of interferences, especially those eluting close to DHEA and epi-androsterone; the GC-C-IRMS measurement which is very demanding in terms of peak purity cannot be performed without this step.

5.3.7.1 Derivatisation

5.3.7.1.1 Oxidation of Glucocorticoids. The analysis of glucocorticoids by GC requires the protection or a modification of the hydroxyl functions.[28] MO-TMS derivatisation was used in several studies[109] involving a ketoprotection with methoxylamine hydrochloride in pyridine followed by a silylation of the hydroxyl functions with BSA (*N*,*O*-bis(trimethylsilyl)acetamide). Good GC behaviour of cortisone and cortisol derivatives was demonstrated but this type of derivative is not ideal for GC-IRMS due to the addition of a significant number of carbon atoms. An alternative method proposed more recently[90] involved the oxidation of all corticoid hydroxylated functions with potassium bichromate in sulfuric acid. This approach presents a double advantage; no carbon atoms are introduced into the chemical structure and, moreover, simplification of the chemical structures is observed. The structure of the parent corticosteroid defines the structure of the oxidised product, for example 17α-hydroxyprogesterone and 11-desoxycortisol are oxidised to 4-androstene-3,17-dione whereas tetrahydrocortisol and tetrahydrocortisone are converted into 5-androstane-3,11,17-trione. Excellent recovery of the oxidation product was obtained in both these cases. In the oxidation of cortisol and cortisone, 30% of the unreacted parent steroid was detected, whereas other target corticoids were totally converted into their oxidised form.[28]

The oxidative step generated a significant amount of by-products and the oxidised glucocorticoids required further purification prior to analysis by GC-C-IRMS. A silica SPE phase was considered because of its capability to trap major oxidised interferences by hydrogen bonding.[28] Nevertheless, in spite of the improvement in selectivity with silica SPE, some remaining interferences could affect the accuracy of the isotopic deviation measurement, particularly for 5*β*-androstane-3,11,17-trione and 4-androstene-3,11,17-trione. A final purification on semi-preparative HPLC is strongly recommended to ensure the chromatographic peak purity of each analyte of interest.[28]

5.3.7.1.2 Acetylation of Steroids. The first derivatisation approach developed in the study of steroid analysis by GC-IRMS relied upon trimethylsilylation using MSTFA/NH_4I/DTE because this reagent was widely used in the field of steroid analysis for the control of drugs in sport and has proven to be robust.[108] The main advantage of the technique was above all the good chromatographic behaviour of the derivatised analytes, a single synthesised product and the high derivatisation yield. The main disadvantage was due to the consequent (up to 6) number of carbons introduced in the molecule disrupting the $^{13}C/^{12}C$ measurement. The steroid $^{13}C/^{12}C$ value is 25% modified with the contribution of introduced carbons during the derivatisation step, according to the formula presented in Figure 5.10.[110]. Thus, the accurate isotopic deviation of steroids could be negatively affected. Moreover, the silylation was much debated for another reason: the stability of the combustion interface. Some authors reported[112] a rapid clogging of the furnace after repeated injection of silylation products and the reagents.

$$\delta^{13}C_{compound}\,(‰) = \frac{n_{derivised\ compound}\,\delta^{13}C_{derivitised\ compound} - n_{derivative\ group}\,\delta^{13}C_{derivative\ group}}{n_{compound}}$$

Figure 5.10 Formula used for the correction of each added carbon after derivatisation (n: number of carbon atoms).

Acetylation presented an alternative and was preferred because only two carbons were introduced on each alcohol function.[90,113,114] In addition, acetylated steroids are stable over weeks in solvents such as cyclohexane, and can be further purified when necessary. Ideally, non-derivatised steroids would have been a better choice for GC-C-IRMS measurements as the steroid $^{13}C/^{12}C$ ratio is not affected; but non-protected steroids clearly show a worse chromatographic behaviour, particularly in the case of oestrogens. The underivatised steroids showed significant increase of peak tailing, peak width and peak asymmetry when compared to acetylated steroids. The derivatisation leads to reduced peak tailing, allowing for more efficient integration and finally better characterisation. It significantly improves the repeatability and the precision of the $^{13}C/^{12}C$ measurement.

5.3.7.2 Chromatography and Isotopic Fractionation

Isotopic fractionation is identified as a major analytical hazard regarding the robustness of IRMS measurement. This phenomenon can occur all along the chromatographic processes from the first SPE to the final GC separation, but the most critical certainly remains HPLC purification stages.[27] Elution profiles of non-derivatised steroids from the dimethylamino-grafted silica (used as a normal phase) or derivatised steroids on C_{18}-grafted silica showed significant modification of the $^{13}C/^{12}C$ ratio from the start of the peak to the end. On the normal phase column dimethylaminopropyl (Figure 5.11), lower $\delta^{13}C_{V-PDB}$ values were observed for DHEA at the beginning of the elution (–36‰ for the acetylated analyte) compared to the end (–33‰ for the acetylated analyte). For 17α-estradiol eluted on the same stationary phase, the highest $\delta^{13}C_{V-PDB}$ values were recorded at the beginning of the peak elution with a significant difference of 4‰ between the start and the end. On the C_{18} stationary phase, higher $\delta^{13}C_{V-PDB}$ values were obtained at the beginning of the elution for both steroids (Figure 5.11). Isotopic deviations of the different fractions exceeded 17‰ for acetylated DHEA from start to end. The data published by Buisson *et al.*[27] corroborated those observed by Kenig *et al.*;[102] in this case, an 18‰ ^{13}C isotopic fractionation across a peak was reported on a reversed phase. As a consequence, the fraction collection has to be definitely considered as a critical step of the purification regarding measurement accuracy. To guarantee a reproducible peak collection for the target steroids, a DAD system is generally used, coupled online to a collector. The collection windows of steroids were monitored at 205 nm for androgens and 280 nm for oestrogens.

5.3.8 Quality Control Samples

To assure the quality of the measurements by GC-IRMS, a number of quality control samples tend to be analysed on a daily basis in the laboratories performing these measurements:

- A spiked urine sample to control the recovery and fractionation.

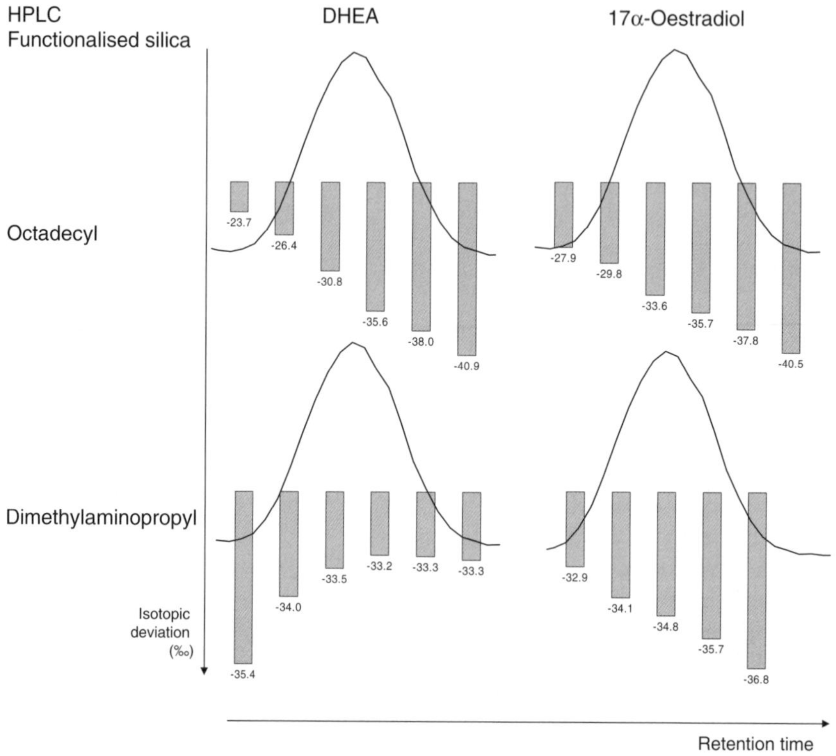

Figure 5.11 Observation of steroid isotopic fractionation on a dimethylaminopropyl-grafted silica ($N(CH_3)_2$) and on an octadecyl-grafted silica (C_{18}) for DHEA and 17α-oestradiol. $\delta^{13}C_{V-PDB}$ values corresponding to acetylated steroids (non-corrected values).

- A blank urine sample to prevent any cross-contamination during the analytical procedure.
- A compliant urine sample as a quality control (QC) point to check for the threshold (see Section 5.3.9) in the sequence (for compliant samples). Urine samples collected from a non-treated animal are used; care is taken to assure that both ERC and metabolites are measurable by GC-C-IRMS.
- A non-compliant urine sample as a quality control (QC) point to check for the threshold (see Section 5.3.9) in the sequence (for non-compliant samples). Urine samples collected from a treated animal are used; care is taken to assure that both ERC and metabolites are measurable by GC-C-IRMS and their difference in terms of isotopic composition is above the threshold of non-compliance (see section 5.3.9).

Other classical precautions such as external standard calibration or additional HPLC fractions to the original target ones could be efficient tools to assure the quality of the measurement.

5.3.9 Definition of a Compliant Threshold in Cattle Urine

In the antidoping world in sport, an official threshold has been set up and published by the World Antidoping Agency[89] " . . . *the results will be reported as consistent with the administration of a steroid when the* $^{13}C/^{12}C$ *value measured for the metabolite(s) differs significantly i.e. by 3 delta units or more from that of the urinary reference steroid chosen* . . . ".

No official criterion regarding Δδ‰ is available in the food safety community, nor are there any scientific papers discussing this issue.

5.3.9.1 Threshold Setting

Depending on the nature of the endogenous steroid suspected to have been administered, the analysed target metabolites are 17α-testosterone and etiocholanolone to monitor testosterone administration and 17α-oestradiol to monitor oestradiol administration while the urinary reference steroids usually used are DHEA and/or 5-androstene-3β,17α-diol.

The results are reported as consistent with the administration of a steroid when the $^{13}C/^{12}C$ value measured for the metabolite(s) differs significantly from that of the ERC.

The threshold can be established on the basis of the measurement of a wide range of compliant urines, for which the mean value and the associated standard deviation of the difference in between ERC and metabolite(s) is calculated.[115] A security factor is applicable to minimise the false positive score.

$$\text{Threshold} = \mu_{\Delta\delta}(\text{metabolite-ERC}) - \mathbf{t}.\sigma_{\Delta\delta\text{metabolite-ERC}},$$

where t corresponds to the value in the Student table for *n* measurements with an α risk set at 1%, μ is the mean value and σ the standard deviation.

5.3.9.2 Threshold Values as Proposed in Cattle

Depending on the target metabolite, etiocholanolone, 17α-testosterone or 17α-oestradiol, Bichon *et al.*[28] proposed thresholds (Figure 5.12) from –2.2 to –2.9‰, respectively. The statistical confidence associated to each threshold is directly linked to the number of the measurements (*n*). For testosterone metabolites, *i.e.* etiocholanolone and 17α-testosterone, a significant number of bovine urine samples have been analysed from animals of different age and gender (calf, heifer, steer, cow, bull and beef). For 17α-oestradiol, measurable concentrations ($> 20\,\mu g\,l^{-1}$) are possible only in pregnant and old cows.

For the official control, samples should be nowadays reported as consistent with the administration of testosterone/oestradiol when the $^{13}C/^{12}C$ value measured for the metabolite(s) differs by three delta units or more from that of the reference steroid chosen.

COMPLIANT THRESHOLD FOR TESTOSTERONE

	Compliant threshold	Compliant threshold
metabolite	etiocholanolone	17α-testosterone
ERC	DHEA	DHEA
average	-0.35	-0.62
standard deviation	0.79	0.92
n measurement	85	38
t (α= 1%)	2.33	2.33
CT = μ + t. σ	**-2.2**	**-2.8**

COMPLIANT THRESHOLD FOR OESTRADIOL

	Compliant threshold
metabolite	17α-oestradiol
ERC	5-androstene-3 β,17α-diol
average	-0.81
standard deviation	0.85
n measurement	21
t (α= 1%)	2.52
CT = μ + t. σ	**-2.9**

t Table (Fisher and Yates, Statistical tables for biological, agricultural and medical research (Oliver and Boyd, Edinburgh)) for a unilateral risk α = 1%

nb val	t	nb val	t	nb val	t
1	31.821	11	2.718	21	2.518
2	6.965	12	2.681	22	2.508
3	4.541	13	2.650	23	2.500
4	3.747	14	2.624	24	2.492
5	3.365	15	2.602	25	2.485
6	3.143	16	2.583	26	2.479
7	2.998	17	2.567	27	2.473
8	2.896	18	2.552	28	2.467
9	2.821	19	2.539	29	2.462
10	2.764	20	2.528	30	2.457
				∞	2.326

Figure 5.12 Proposal of compliant threshold for the control of testosterone and oestradiol abuse in cattle.

5.3.10 Application to Natural Steroids Monitoring by GC-C-IRMS

5.3.10.1 Case study – Testosterone and/or Androstenedione Administration

The urinary $^{13}C/^{12}C$ profiles of three metabolites (*i.e.* etiocholanolone, 17α-testosterone and 5α-androstane-3β,17α-diol) determined by GC-C-IRMS are

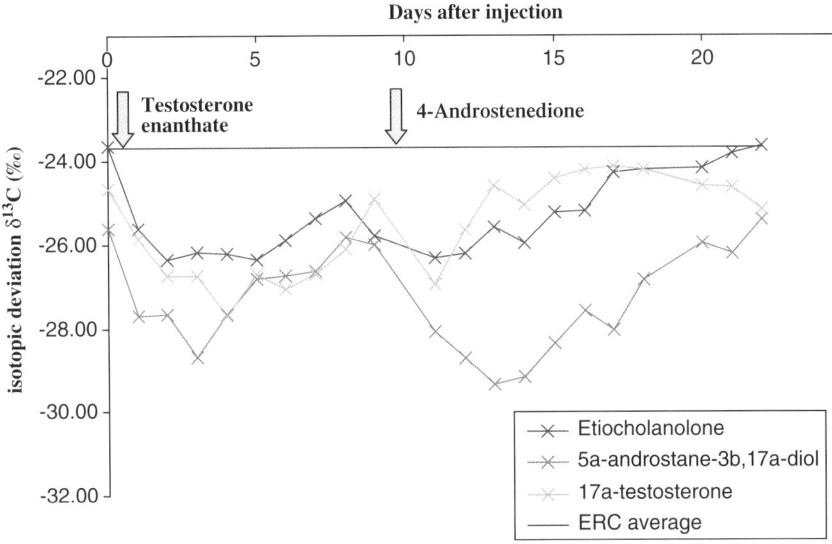

Figure 5.13 Isotopic deviation measurement (expressed in $\delta^{13}C_{VPDB}$ in ‰) of ERC (DHEA and 5-androstene-3β,17α-diol) and metabolites of testosterone and 4-androstenedione (etiocholanolone, 5α-androstane-3β,17α-diol and 17α-testosterone) during 10 days after testosterone enanthate administration (250 mg, IM injection) and 12 days after 4-androstenedione injection (100 mg).

shown in Figure 5.13 following the administration of testosterone enanthate and 4-androstenedione to a bovine.[27] The average value of the ERC (–23.7±0.6‰ [$n = 43$]) demonstrated the robustness of the analytical methodology and the relative homogeneity of the endogenous steroid isotopic deviation (consequence of the diet, mainly based on hay). Figure 5.13 illustrates the depletion of the metabolite isotopic deviation following injection of 17α-testosterone and shows a significant difference in between ERC and metabolites (etiocholanone or 5α-androstane-3β,17α-diol) for over one week after injection.

5.3.10.2 Case Study – Oestradiol

Buisson *et al.*[27] studied the efficiency of such an approach to demonstrate oestradiol administration to cattle after estradiol valerate injection. The $\delta^{13}C_{VPDB}$-values of both ERC (*i.e.* DHEA and 5-androstene-3β,17α-diol) and the main oestradiol metabolite (17α-oestradiol) were measured in urine samples collected in different animals, treated *versus* non-treated, gender (male, female *versus* castrated), age (sexually mature and immature) and feedings (grass or maize). The metabolite – 17α-oestradiol – was found to be difficult to measure in some samples especially in untreated animals and/or pregnant cows. The ERC $^{13}C/^{12}C$ ratio was not affected by the oestradiol treatment and found very repeatable one animal to another when feed remained constant. For

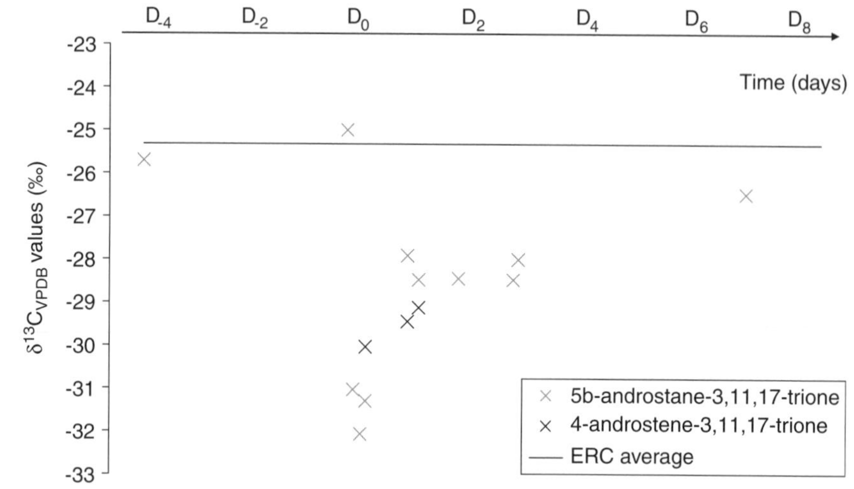

Figure 5.14 $\delta^{13}C_{V-PDB}$ values of DHEA, 5β-androstane-3,11,17-trione and 4-androstene-3,11,17-trione in urine samples between D_{-4} and D_8; D_0 corresponds to the day of injection.

oestrogens, the $\delta^{13}C_{VPDB}$ values of 17α-estradiol in an untreated pregnant cow fed with maize were found to be –17.8‰. The Δδ‰, *i.e.* the difference between the $\delta^{13}C_{VPDB}$ value of the ERC and 17α-oestradiol after treatment is higher than 14‰. This difference is substantial and far above the measurement uncertainty and unambiguously allowed the differentiation in between treated and non-treated animals.

5.3.10.3 Case Study – Cortisol

One of the main difficulties regarding the control of natural corticosteroids remains their rapid urinary elimination from the body so that the detection of residues does not exceed in general one week after administration.[28] Because of their relatively high polarity, their separation by gas chromatography needs a derivatisation of the molecule. Pujos *et al.*[99] and Bichon *et al.*[28] developed a strategy based on the oxidation (Figure 5.8) of corticosteroids, leading to the synthesis of 4-androstene-3,11,17-trione for cortisol/cortisone and 5*β*-androstane-3,11,17-trione for their metabolites. After administration of cortisol, a comparable and very fast decrease of the isotopic deviation was observed for the two target analytes corresponding to the metabolites (Figure 5.14); values were depleted down to –30‰ and –32‰, respectively. The diet of the animal was kept unchanged during the experiment; the isotopic deviation of the ERC was thus observed constant (–25.3±0.4‰). The corresponding Δ(*δ*‰) (ERC-metabolites) after treatment showed difference up to 7‰ immediately after administration and 3‰ until three days after. Up to now, no threshold level has been proposed. Additional experiments are necessary.

5.4 Conclusion

Developments in chromatographic and mass spectrometric techniques are providing the analyst in the field of drug residue analysis in livestock with some powerful tools to address the emerging challenges arising from the development of new and more potent drugs, administration of cocktails and, also, the challenges presented from the advances in the biotechnology industry in the production of protein-based drugs. Of particular interest is the development of multi-residue screening procedures and their validation according to EC recommended guidelines. Methods have been developed through the use of high-resolution chromatographic techniques coupled with high-resolution mass spectrometry that allow for over 100 drugs to be covered in a single analysis. It will be interesting to note the impact this approach has upon the field of veterinary drug residue analysis in the forthcoming years and the policies that emerge to address the assay validation issues.

The IRMS methodology is nowadays becoming more and more popular for confirmation purposes regarding the determination of natural steroid origin. The method is officially applied in antidoping and food safety to control testosterone, oestradiol and research studies are currently running for nandrolone, boldenone and cortisol. An official threshold is now proposed for testosterone and oestradiol in cattle, but it has not been accepted officially by the European authorities. Improvements are expected in the sensitivity of the instrument to facilitate the $^{13}C/^{12}C$ measurement of other natural steroids and to allow the technique to be applied to other biological matrices such as tissue samples. Extension to other isotopes would be beneficial to ensure the unambiguous character of the conclusion; $^{2}H/^{1}H$ is probably the next item. A possible further development would be the GCxGC approach to improve the chromatographic separation, as it remains a critical limitation in the robust determination of isotope composition. Finally, for other classes of compounds such as growth hormones (somatotropine), new coupling such as LC-IRMS would be beneficial to the domain.

The application of "omics" technologies to detect biomarkers of drug administration is an exciting novel approach being applied to drug detection in sport and also in veterinary drug residue analysis. The initial publications in this area clearly demonstrate the potential of applying proteomics, transcriptomics or metabolomics to address the challenges to the industry and demonstrate the administration of a substance or substances with a particular pharmacological action. The additional challenge for the analyst may then be to identify the particular substance administered or for the authorities to define policies for the use of such approaches to control the misuse of drugs.

Acknowledgements

The authors acknowledge the contribution of Dr Richard Rodenburg, formerly TNO and now at the Radboud University, Nijmegen, to the summary discussion of "omics" technologies in Section 5.3.2 of this chapter.

References

1. D. Courtheyn, B. Le Bizec, G. Brambilla, H. F. De Brabender, E. Cobbaert, M. Van De Wiele, J. Vercammen and K. De Wasch, *Anal. Chim. Acta*, 2002, **473**(1), 71–82.
2. H. F. De Brabender, H. Noppe, K. Verheyden, J. Vanden Bussche, K. Wille, L. Okerman, L. Vanhaecke, W. Reybroeck, S. Ooghe and S. Croubels, *J. Chrom. A*, in press, *CHROMA-349789*.
3. A. A. M. Stolker and U. A. Th. Brinkman, *J. Chrom. A*, 2005, **1067**, 15–53.
4. D. H. Catlin, M. H. Sekera, B. D. Ahrens, B. Starcevic, Y. -C. Chang and C. K. Hatton, *Rapid Commun. Mass Spectrom.*, 2004, **18**, 1245–1249.
5. D. H. Catlin, K. D. Fitch and A. Ljungqvist, *J. Internal Med.*, 2008, **264**, 99–114.
6. A. Baoutina, I. E. Alexansder, J. E. J. Rasko and K. R. Emslie, *J. Gene Med.*, 2008, **10**, 3–20.
7. W. A. Storer, D. L. Thompson Jr, K. R. Bondioli, R. A. Godke, A. S. Khan, P. A. Brown and R. Draghia-Akli, *J. Equine Vet. Sci.*, 2008, **28**(3), 149–155.
8. P. Marchand, B. le Bizec, C. Gade, F. Monteau and F. André, *J. Chromatogr. A*, 2000, **867**, 219.
9. F. Courant, J. -P. Antignac, D. Maume, F. Monteau, F. Andre and B. L. Bizec, *Food Addit. Contam. A.*, 2007, **24**, 1358.
10. J. -P. Antignac, P. Marchand, B. Le Bizec and F. Andre, *J. Chromatogr. B Analytical Technologies in the Biomedical and Life Sciences*, 2002, **774**, 59.
11. H. Noppe, B. Le Bizec, K. Verheyden and H. F. De Brabander, *Anal. Chim. Acta*, 2008, **611**, 1.
12. L. A. Van Ginkel, R. W. Stephany, H. J. Van Rossum, H. Van Blitterswijk, P. W. Zoontjes and R. C. M. Hooijschuur, *J. Chromatogr. Biomed. Appl.*, 1989, **489**(1), 95–104.
13. T. P. Samuels, A. Nedderman, M. A. Seymour and E. Houghton, *The Analyst*, 1998, **123**, 2401–2404.
14. J. Segura, S. Pichini, S. H. Peng and X. de la Torre, *Forensic Sci. Int.*, 2000, **107**, 347.
15. P. Kintz, V. Cirimele, H. Sachs, T. Jeanneau and B. Ludes, *Forensic Sci. Int.*, 1999, **101**, 209.
16. H. Hooijerink, A. Lommen, P. P. J. Mulder, J. A. van Rhijn and M. H. W. F. Nielen, *Anal. Chim. Acta EURORESIDUE V,* Noordwijkerhout, The Netherlands, 10–12 May 2004, 2005, **529**, 167.
17. L. Rambaud, F. Monteau, Y. Deceuninck, E. Bichon, F. André and B. Le Bizec, *Anal. Chim. Acta Papers Presented at the 5th International Symposium on Hormone and Veterinary Drug Residue Analysis, Papers Presented at the 5th International Symposium on Hormone and Veterinary drug Residue Analysis*, 2007, **586**, 93.
18. L. Rambaud, E. Bichon, N. Cesbron, F. André and B. L. Bizec, *Anal. Chim. Acta*, 2005, **532**, 165.

19. B. Le Bizec, F. Courant, I. Gaudin, E. Bichon, B. Destrez, R. Schilt, R. Draisci, F. Monteau and F. André, *Steroids*, 2006, **71**, 1078.
20. M. A. Popot, E. Houghton, A. Ginn, M. Jones, P. Teale, T. Samuels, V. Lassourd, N. Dunnett, D. A. Cowan, Y. Bonnaire and P. L. Toutain, *Equine Vet. J.*, 1997, **29**, 226.
21. Sanco/D3/ABR/ca, D/430638, Note to the members of the standing committee on the food chain and animal health, boldenone in calves, update, 2005.
22. T. Furuta, N. Eguchi, H. Shibasaki and Y. Kasuya, *J. Chromatogr. B Biomed. Sci. Appl.*, 2000, **738**, 119.
23. P. M. Mason, I. Gilmour, C. Pillinger, S. E. Hall, E. Houghton and M. A. Seymour, *Analyst*, 1998, **123**, 2405.
24. V. Ferchaud, B. Le Bizec, F. Monteau and F. André, *Rapid Comm. Mass Spectrom.*, 2000, **14**, 652.
25. V. Ferchaud, B. Le Bizec, F. Monteau and F. André, *Analyst*, 1998, **123**, 2617.
26. M. Hebestreit, U. Flenker, C. Buisson, F. Andre, B. Le Bizec, H. Fry, M. Lang, A. P. Weigert, K. Heinrich, S. Hird and W. Schanzer, *J. Agri. Food Chem.*, 2006, **54**, 2850.
27. C. Buisson, M. Hebestreit, A. P. Weigert, K. Heinrich, H. Fry, U. Flenker, S. Banneke, S. Prevost, F. Andre, W. Schaenzer, E. Houghton and B. Le Bizec, *J. Chromatogr. A*, 2005, **1093**, 69.
28. E. Bichon, *Rapid Comm. Mass Spectrom.*, 2007, **21**, 2613.
29. H. F. De Brabander, B. Le Bizec, G. Pinel, J. P. Antignac, K. Verheyden, V. Mortier, D. Courtehyn and H. Noppe, *J. Mass Spectrom.*, 2007, **42**(8), 983.
30. M. W. Nielen, M. C. van Engelen, R. Zuiderent and R. Ramaker, *Anal. Chim. Acta*, 2007, **586**(1–2), 122.
31. E. van der Heeft, Y. J. Bolck, B. Beumer, A. W. Nijrolder, A. A. Stolker and M. W. Nielen, *J. Am. Soc. Mass Spectrom.*, 2009, **20**(3), 451.
32. E. van der Heeft, P. Zomer, L. A. M. Stolker and M. W. F. Nielen, in *Proceedings of the EuroResidue VI Conference*, ed. L. A. Van Ginkel and A. A. Bergwerff, 2008, vol. 2, p. 565.
33. F. Monteau, L. Rambaud, Y. Deceuninck, E. Bichon, G. Pinel, M.-H. Le breton, J.-P. Antignac, D. Maume and B. Le Bizec, in *Proceedings of the EuroResidue VI Conference*, ed. L. A. Van Ginkel and A. A. Bergwerff, 2008, vol. 2, p. 571.
34. Y. Deceuninck, E. Bichon, B. Destrez, L. Ranbaud, F. Courant, F. Monteau, J.-P. Antignac and B. Le Bizec, in *Proceedings of the EuroResidue VI Conference*, ed. L. A. Van Ginkel and A. A. Bergwerff, 2008, vol. 2, p. 581.
35. M.-H. Le Breton, S. Roulet-Rochereau, G. Pinel, T. Goldman, J.-M. Diserens and B. Le Bizec, in *Proceedings of the EuroResidue VI Conference*, ed. L. A. Van Ginkel and A. A. Bergwerff, 2008, vol. 1, p. 253.

36. M. -H. Le Breton, S. Roulet-Rochereau, G. Pinel, L. Bailly-Chouriberry, G. Rychen, S. Jurjanz, T. Goldman and B. Le Bizec, *Rapid Comm. Mass Spectrom.*, 2008, **22**, 3130.
37. F. Guan, C. E. Uboh, L. R. Soma, E. Birks, J. Chen, J. Mitchell, Y. You, J. Rudy, F. Xu, X. Li and G. Mbuy, *Anal. Chem.*, 2005, **79**(12), 4627.
38. S. Bobin, M. -A. Popot, Y. Bonnaire and J. C. Tabet, *Analyst*, 2001, **126**, 1996–2001.
39. N. L. Anderson, N. G. Anderson, L. R. Haines and D. B. Hardie, *et al.*, *J. Proteome Res.*, 2004, **3**, 235–244.
40. C. Barton, P. Beck, R. Kay, P. Teale, and J. Roberts, *Proteomics*, 2009, accepted for publication.
41. F. Kay, B. Gregory, P. B. Grace and S. Pleasance, *Rapid Comm. Mass Spectrom.*, 2007, **21**, 2585–2593.
42. A. Kaufmann, P. Butcher, K. Maden and M. Widmer, *J. Chromatogr. A*, 2008, **1194**, 66.
43. S. Abuin, R. Companyo, F. Centrich, A. Rubies and M. D. Prat, *J. Chromatogr. A*, 2008, **1207**, 17.
44. B. Kasprzyk-Hordern, R. M. Dinsdale and A. J. Gewy, *J. Chromatogr. A*, 2007, **1161**(1–2), 132.
45. M. Petrovic, M. Gros and D. Barcelo, *J. Chromatogr. A*, 2006, **1124**(1–2), 68.
46. B. Shao, X. Jia, Y. Wu, J. Hu, X. Tu and J. Zhang, *Rapid Common. Mass Spectrom.*, 2007, **21**(21), 3487.
47. A. Kayfmann, P. Butcher, K. Maden and M. Widmer, *Anal. Chim. Acta*, 2007, **5869**(1–2), 13.
48. A. A. Stolker, P. Ritgers, E. Oosterink, J. J. Lasaroms, R. J. Peters, J. A. Rhijn and M. W. Nielen, *Anal. Bioanal. Chem.*, 2008, **391**(6), 2309.
49. M. W. Nielen, J. J. Lasaroms, M. L. Essers, J. E. Oosterink, T. Meijer, M. B. Sanders, T. Zuidema and A. A. Stolker, *Anal. Bioanal. Chem.*, 2008, **391**(1), 199.
50. I. Riedmaier, C. Becker, W. Pfaffl, H. H. D. Meyer, J. *Chromatrogr. A* (*CHROMA-349755*), in press.
51. The International Genome Sequencing Consortium, *Nature*, 2004, **431**, 931–945.
52. J. van der Greef, in *Abstracts, 4th International Symposium on Hormone and Veterinary Drug Residue Analysis,* Antwerp, Belgium, 2002.
53. R. Dijkmans, D. Groot Kormeling, K. Sultan, B. Jongerius, B. Blankvoort, R. Schilt, R. Witkamp, H. Haagsman and A. Bergwerff, in *Abstracts, 4th International Symposium on Hormone and Veterinary Drug Residue analysis,* Antwerp, Belgium, 2002.
54. E. Kamanga-Sollo, M. S. Pampusch, G. Xi, M. E. White, M. R. Hathaway and W. R. Dayton, *J. Cell Physiol.*, 2004, **201**(2), 181–189.
55. M. S. Pampusch, B. J. Johnson, M. E. White, M. R. Hathaway, J. D. Dunn, A. T. Waylan and W. R. Dayton, *J. Anim. Sci.*, 2003, **81**(11), 2733–2740.

56. M. S. Pampusch, M. E. White, M. R. Hathaway, T. J. Baxa, K. Y. Chung, S. L. Parr, B. J. Johnson, W. J. Weber and W. R. Dayton, *J. Anim. Sci.*, 2008, **86**(12), 3418–3423, Epub 2008 Aug 1.
57. J. Honglin, W. Ying, W. Miaozong, G. Zhiliang, J. F. Stuart and D. Torres-Diaz, *Endocrinology*, 2007, **148**(7), 3307–3315.
58. M. W. Pfaffl, I. G. Lange, A. Daxenberger and H. H. Meyer, *APMIS*, 2001, **109**(5), 345–355.
59. M. Reiter, V. M. Walf, A. Christians, M. W. Pfaffl and H. H. D. Meyer, *Anal. Chim. Acta*, 2007, **586**(1–2), 73–81.
60. M. W. Pfaffl, A. Daxenberger, M. Hageleit and H. H. Meyer, *J. Vet. Med. A Physiol. Pathol. Clin. Med.*, 2002, **49**(2), 57–64.
61. E. R. Mardis, *Trends Genet.*, 2008, **24**(3), 133–141.
62. C. Bendixen, J. Hedegaard and P. Horn, *Meat Science*, 2005, **71**(1), 128–137.
63. J. C. W. Rijk, A. C. M. Peijnenburg, H. Baykus, M. J. Groot, J. M. Van Hende and M. W. Nielen, in *Proceedings of the EuroResidue VI Conference,* Egmond aan Zee, The Netherlands, 2008, 171–174.
64. L. Carraro, S. Ferraresso, B. Cardazzo, A. N. Mininni, C. Montesissa and M. Castagnaro, in *Proceedings of the EuroResidue VI Conference,* Egmond aan Zee, The Netherlands, 2008, 1385–1390.
65. G. Gardini, P. Del Boccio, S. Colombatto, G. Testore, D. Corpillo, C. Di Ilio, A. Urbani and C. Nebbia, *Proteomics*, 2006, **6**, 2813–2822.
66. R. I. G. Holt and P. H. Sonksen, *J. Pharmacol.*, 2008, **154**, 542–556.
67. R. Renaville, S. Massart, G. Logany, A. Devolder, M. Sneyers, M. Marlier, M. Severlin, A. Burny and D. Portetelle, *Anim. Prod.*, 1994, **59**, 189–196.
68. B. J. Johnson, M. R. Hathaway, P. T. Anderson, J. C. Meiske and W. R. Dayton, *J. Anim. Sci.*, 1996, **74**(2), 372–379.
69. R. Renaville, M. Hammadi and D. Portetelle, *Domest. Anim. Endocrinol.*, 2002, **23**(1–2), 351–60.
70. R. Draisci, C. Montesissa, B. Santamaria, C. D'Ambrosio, G. Ferretti, R. Merlanti, C. Ferranti, M. De Liguoro, C. Cartoni, E. Pistarino, L. Ferrara, M. Tiso, A. Scaloni and M. E. Cosulich, *Proteomics*, 2007, **7**(17), 3184–3193.
71. M. H. Mooney, C. Situ, G. Cacciatore, T. Hutchinson, C. Elliott and A. A. Bergwerff, *Biomarkers*, 2008, **13**(3), 246–256.
72. M. Mooney, A. A. van Meeuwen, C. Situ, G. Cacciatore, P. Delahaut, E. De Pauw, A. A. Bergwerff and C. Elliott, *Proc. EuroResidue VI*, May, 2008, p. 159–163.
73. G. Biancotto, I. Andrighetto, R. Angeletti, G. Pozza, M. Vascellari, F. Mutinelli, L. Poppi, R. Stella, G. Arrigoni, M. C. Corgato, M. Krogh and P. James, *Proc. EuroResidue VI, May*, 2008, p. 1285–1291.
74. G. Cacciatore, S. W. Eisenberg, C. Situ, M. H. Mooney, P. Delahaut, S. Klarenbeek, A. C. Huet, A. A. Bergwerff and C. T. Elliott, *Anal. Chim. Acta*, 2009, **637**(1–2), 351–359.

75. J. van Meeuwen, M. H Mooney, C. Charlier, P. Delahaut, J. Buijs, C. T. Elliott and A. A. Bergwerff, *Proc. EuroResidue VI*, 2008, p. 175–180.
76. M. Bedair and L. W. Sumner, *Trends Anal. Chem.*, 2008, **27**, 238.
77. M. J. Gibney, M. Walsh, L. Brennan, H. M Roche, B. German and B. van Ommen, *Am.J. Clin. Nutr.*, 2005, **82**, 497.
78. A. Lommen, G. van der Weg, M. C. van Engelen, G. Bor, L. A. Hoogenboom and M. W. Nielen, *Anal. Chim. Acta*, 2007, **584**(1), 43–49, Epub 2006 Nov 11.
79. A. Lommen, H. H. Heskamp, M. K. Van der Lee, G. Van der Weg, H. G. J Mol and M. W. Nielen, *Proc. EuroResidue VI*, 2008, p. 1245–1249.
80. M.-E. Dumas, C. Canlet, J. Vercauteren, F. André and A. Paris, *J. Proteome Res.*, 2005, **4**, 1493.
81. J.-P. Antignac, G. Pinel, E. Bichon, F. Monteau, F. Courant, F. Kieken, B. Destrez, B. Le Bizec, *Proc. EuroResidue VI*, 2008, p. 165–170.
82. R. T. Cunningham, M. H. Mooney, X. L. Xia, S. Crooks, D. Matthews, M. O'Keeffe, K. Li and C. T. Elliott, *Anal. Chem.*, 2009, Epub ahead of print.
83. F. Kieken, G. Pinel, J.-P. Antignac, M.-A. Popot, Y. Bonnaire and B. Le Bizec, *Proc. EuroResidue VI*, 2008, p. 511–515.
84. G. D. Farquhar, J. R. Ehleringer and K. T. Hubick, *Annu. Rev. Plant Physiol. Plant Mol. Biol.*, 1989, **40**, 503.
85. A. Kleemann and H. J. Roth, *Thieme,* Stuttgart, 1983.
86. H. Craig, *Geochim. Cosmochim. Acta*, 1957, **12**, 133.
87. COMMISSION REGULATION (EC) No 440/2003 of 10 March 2003 amending Regulation (EEC) No 2676/90 determining Community methods for the analysis of wines (L 66/15).
88. M. Becchi, R. Aguilera, Y. Farizon, M.-M. Flament, H. Casabianca and P. James, *Rapid Comm. Mass Spectrom.*, 1994, **8**, 304.
89. WADA TD2004EAAS v1.0, 30 May 2004, Effective Date: 13 August 2004.
90. E. Pujos, M. M. Flament-Waton, P. Goetinck and M. F. Grenier-Loustalot, *Analytical and Bioanalytical Chemistry*, 2004, **380**, 524.
91. M. F. Grenier-Loustalot, *Science & Sports Recherche et lutte contre le dopage – Colloque organisé par le Conseil de prévention et de lutte contre le dopage*, Paris, 20 Janvier 2005, 2005, **20**, 208.
92. E. Bourgogne, V. Herrou, J.-C. Mathurin, M. Becchi and J. de Ceaurriz, *Rapid Comm. Mass Spectrom.*, 2000, **14**, 2343.
93. Y. Duan, Q. B. Wen and B. J. Luo, *China Organic Geochemistry*, 1997, **27**, 583.
94. W. Meier-Augenstein, *Anal. Chim. Acta*, 2002, **465**, 63.
95. E. Jamin, J. González, G. Remaud, N. Naulet, G. G. Martin, D. Weber, A. Rossmann and H. L. Schmidt, *Anal. Chim. Acta*, 1997, **347**, 359.
96. A. Albertino, A. Barge, G. Cravotto, L. Genzini, R. Gobetto and M. Vincenti, *Food Chem.*, 2009, **112**, 715.

97. D. Fabbri, I. Vassura, C.-G. Sun, C. E. Snape, C. McRae and A. E. Fallick, *Mar. Chem.*, 2003, **84**, 123.
98. W. Vetter, *Environment International Proceedings of the 1st Conference of the UK Network on Persistent Organic Pollutants (POPs)* 29th and 30th March 2006, University of Birmingham, UK, 2008, **34**, 357.
99. L. M. Reid, C. P. O'Donnell and G. Downey, *Trends Food Sci. Tech.*, 2006, **17**, 344.
100. G. Balizs, A. Jainz and P. Horvatovich, *J Chromatogr. A*, 2005, **1067**, 323.
101. K. Mastovska and S. Lehotay, *J. Chromatogr. A*, 2003, **1000**, 153.
102. F. Kenig, B. N. Popp and R. E. Summons, *Org. Geochem.*, 2000, **31**, 1087.
103. R. J. Caimi and J. T. Brenna, *J. Chromatogr. A*, 1997, **757**, 307.
104. E. Venturelli, A. Cavalleri and G. Secreto, *J. Chromatogr. B Biomed. Sci. Appl.*, 1995, **671**, 363.
105. C. H. L. Shackleton, *J. Chromatogr. B Biomed. Sci. Appl.*, 1986, **379**, 91.
106. J. P. Antignac, A. Brosseaud, I. Gaudin, F. André and B. Le Bizec, *Steroids*, 2005, **70**, 205.
107. P. Vestergaard, *Acta Endocrinol. Suppl.*, 1978, **217**, 76.
108. K. M. Lewis and R. D. Archer, *Steroids*, 1979, **34**, 485.
109. S. Prévost, T. Nicol, F. Monteau, F. André and B. Le Bizec, *Rapid Comm. Mass Spectrom.*, 2001, **15**, 2509.
110. H. Shibasaki, T. Furuta and Y. Kasuya, *J. Chromatogr. Biomed. Appl.*, 1992, **579**, 193.
111. G. Rieley, *Analyst*, 1994, **119**, 915.
112. W. Meier-Augenstein, in *Handbook of Stable Isotope Analytical Techniques*, ed. P. A. D. Groot, Elsevier, Amsterdam, 2004, p. 153.
113. R. Aguilera, M. Becchi, C. Grenot, H. Casabianca and C. K. Hatton, *J. Chromatogr. B Biomed. Sci. Appl.*, 1996, **687**, 43.
114. R. Aguilera, M. Becchi, H. Casabianca, C. K. Hatton, D. H. Catlin and B. Starcevic, *J. Mass Spectrom.*, 1996, **31**, 169.
115. E. Bichon, F. Kieken, N. Cesbron, F. Monteau, S. Prevost, B. Le Bizec, *Proceeding presented for Euroresidue VIth Conference,* Egmond aan Zee, The Netherlands, 2008.

Subject Index

Italic Page Numbers refer to Tables and Figures.